Modeling Remaining Useful Life Dynamics in Reliability Engineering

This book applies traditional reliability engineering methods to prognostics and health management (PHM), looking at remaining useful life (RUL) and its dynamics, to enable engineers to effectively and accurately predict machinery and systems useful lifespan. One of the key tools used in defining and implementing predictive maintenance policies is the RUL indicator. However, it is essential to account for the uncertainty inherent to the RUL, as otherwise predictive maintenance strategies can be incorrect. This can cause high costs or, alternatively, inappropriate decisions. Methods used to estimate RUL are numerous and diverse and, broadly speaking, fall into three categories: model-based, data-driven, or hybrid, which uses both. The author starts by building on established theory and looks at traditional reliability engineering methods through their relation to PHM requirements and presents the concept of RUL loss rate. Following on from this, the author presents an innovative general method for defining a nonlinear transformation enabling the mean residual life to become a linear function of time. He applies this method to frequently encountered time-to-failure distributions, such as Weibull and gamma, and degradation processes. Latest research results, including the author's (some of which were previously unpublished), are drawn upon and combined with very classical work. Statistical estimation techniques are then presented to estimate RUL from field data, and risk-based methods for maintenance optimization are described, including the use of RUL dynamics for predictive maintenance.

The book ends with suggestions for future research, including links with machine learning and deep learning.

The theory is illustrated by industrial examples. Each chapter is followed by a series of exercises.

FEATURES

- Provides both practical and theoretical background of RUL
- Describes how the uncertainty of RUL can be related to RUL loss rate
- Provides new insights into time-to-failure distributions
- Offers tools for predictive maintenance

This book will be of interest to engineers, researchers and students in reliability engineering, prognostics and health management, and maintenance management.

Modeling Remaining Useful Life Dynamics in Reliability Engineering

Pierre Dersin

CRC Press
Taylor & Francis Group
Boca Raton London New York

CRC Press is an imprint of the
Taylor & Francis Group, an **informa** business

First edition published 2023
by CRC Press
6000 Broken Sound Parkway NW, Suite 300, Boca Raton, FL 33487-2742

and by CRC Press
4 Park Square, Milton Park, Abingdon, Oxon, OX14 4RN

CRC Press is an imprint of Taylor & Francis Group, LLC

© 2023 Pierre Dersin

ISBN: 978-1-032-16859-3 (hbk)
ISBN: 978-1-032-16864-7 (pbk)
ISBN: 978-1-003-25068-5 (ebk)

DOI: 10.1201/9781003250685

Typeset in Times
by Newgen Publishing UK

Dedication

To my wife Katia

Contents

Preface

Over many years in industry (mainly in railways but also in electric power generation and transmission), I have had the opportunity to practice reliability engineering and, later on, was involved with the still emerging discipline of prognostics and health management. I have kept wondering why those two communities are still, in many cases, separate, probably for organizational and historical reasons.

Therefore, areas at the confluence of the two disciplines have stirred my interest.

The study of remaining useful life (RUL) is one of those. While physics of failures, and machine learning, are essential tool kits to address that topic, I think it is also useful to keep in mind some classical reliability engineering approaches, going back to the fundamentals, that shed light on the subject—especially the dynamics and the inherent uncertainty.

This is what I tried to do through some publications, and discussions which they brought about led me to believe that there might be substance for a book that would be of use to practitioners and researchers alike. The former will probably identify new applications, while the latter are likely to pose new questions and expand considerably on what I am presenting here.

The book is meant for practitioners, researchers, teachers, and graduate students. Sufficient prerequisites are a college level calculus background and a knowledge of the basics of probability theory and statistics.

Acknowledgments

The author is particularly grateful to Dr. Roberto Rocchetta (SUPSI and formerly TU Eindhoven), for his meticulous expert review of the manuscript and for providing material for the case studies in Chapter 8.

He also thanks warmly Prof. Alessandro Di Bucchianico (Applied Mathematics Department, TU Eindhoven, the Netherlands) for inviting him to spend a week at TU Eindhoven and having stimulating and intense discussions on the subject.

Colleagues at Luleå University of Technology's Operations and Maintenance Engineering Division, in particular Prof. Uday Kumar and Prof. Ramin Karim; and former Alstom colleagues, have contributed indirectly to the emergence of this book, through our productive collaboration. Gratefully acknowledged is also the fruitful collaboration with Prof. Olga Fink (ETH-Z/ EPFL, Switzerland) in PHM over the years, and with Prof. Alessandro Birolini (ETH-Z Emeritus) in Reliability Engineering.

Others who are too numerous to mention here, have also been a source of inspiration.

Last but not least, heartfelt thanks to Prof. Vasiliy Krivtsov (University of Maryland and Ford Motor Co.) and Prof. Joel Nachlas (University of Virginia) for encouraging me to start this book.

Author

Pierre Dersin, PhD, earned a master's in operations research and a PhD in electrical engineering at the Massachusetts Institute of Technology (MIT). He also holds math and electrical engineering degrees from the Université Libre de Bruxelles (Belgium).

Since 2019, he has been Adjunct Professor at Luleå University of Technology (Sweden) in the Operations and Maintenance Engineering Division.

In January 2022, he founded a small consulting company, Eumetry sas, in Louveciennes, France, in the fields of RAMS, PHM, and AI, just after retiring from ALSTOM, where he had spent more than 30 years.

With Alstom, he was RAM (reliability-availability-maintainability) Director from 2007 to 2021 and founded the RAM Center of Excellence. In 2015, he launched the predictive maintenance activity and became PHM (prognostics and health management) Director of ALSTOM Digital Mobility and then ALSTOM Digital and Integrated Systems, St-Ouen, France.

Prior to joining Alstom, he worked in the United States on the reliability of large electric power networks, as part of the Large Scale System Effectiveness Analysis Program sponsored by the US Department of Energy, from MIT and Systems Control Inc., and later, with FABRICOM (Suez Group), on fault detection and diagnostics in industrial systems.

He has contributed a number of communications and publications in scientific conferences and journals in the fields of RAMS, PHM, AI, automatic control and electric power systems, including Engineering Applications of AI, IEEE Transactions on Automatic Control, IEEE Transactions on Power Apparatus and Systems, ESREL, RAMS Symposia, French Lambda-Mu Symposia, the 2012 IEEE-PHM Conference, the 2014 European Conference of the PHM Society (keynote speaker), and WSC 2013.

He serves on the IEEE Reliability Society AdCom and the IEEE Digital Reality Initiative, and chairs the IEEE Reliability Society Technical Committee on Systems of Systems. He is a contributor of four chapters in the *Handbook of RAMS in Railways: Theory and Practice* (CRC Press, Taylor & Francis, 2018), including a chapter on "PHM in Railways". In January 2020, he was awarded the Alan O. Plait Award for the best tutorial at the RAMS conference, "Designing for Availability in Systems, and Systems of Systems".

His main research interests focus on the confluence between RAMS and PHM as well as complex systems resilience and asset management.

1 Introduction

Asset management (Campbell 2022, ISO 2014), and in particular maintenance, plays a fundamental role in keeping the key critical systems of today's complex world operating smoothly and cost-effectively. Accordingly, more than ever, industry and governments pay close attention to optimizing maintenance and asset management policies through sophisticated health management strategies made possible by today's information and communication technology.

Indeed the last two decades have witnessed the triumph of the digital transformation and, in particular, the "Internet of things" which, through the combination of cost-effective sensors and efficient communication infrastructure, allows many industrial items of equipment to communicate in real-time physical magnitudes that can be used to estimate their health condition. In a parallel with human health, let one think for instance of the wonders that can be accomplished with the simple connected watch used by sport fans.

This evolution, combined with the fast-paced development of advanced data processing algorithms ("analytics"), whose implementation is enabled by powerful parallel processing architectures such as graphics processing units, has spurred the emergence of a discipline named Prognostics and Health Management (PHM).

"PHM" refers to a set of methodologies that have been developed since the turn of the century to support predictive maintenance, i.e., maintenance strategies based on the estimated condition of an asset and on the predicted future evolution of that condition (Gouriveau 2016). PHM consists of several steps: detection, diagnostics, and prognostics. Ultimately it is useful only if it supports decision-making in asset management, i.e. the decisions about when to maintain a particular asset, when to let it run to failure, and when to replace it or to retire it. Detection deals with the discovery of anomalous conditions, which eventually, if not acted upon, will lead to a failure, i.e. (as defined by the IEC standard on dependability vocabulary (IEC 2015), "the loss of the ability to perform as required". Diagnostics focuses primarily on the identification of the type, location, and if possible the root cause, of an anomaly. Both detection and diagnostics are the subject of abundant literature (Jolfaei 2022) and will not be addressed here.

DOI: 10.1201/9781003250685-1

The focus of this work will be on prognostics, which is a key enabler of predictive maintenance. Predictive maintenance is a refinement of condition-based maintenance: it relies on the prediction of "remaining useful life" (RUL), i.e., the time that remains until a failure takes place, in the absence of preventive maintenance intervention.

A failure is defined as the loss of a function (IEC 2015) and corresponds to a transition from a normal state to a fault state or an unacceptable operational condition. A failure time can be characterized, for instance, as the first instant at which a given degradation threshold is crossed. Predictive maintenance aims at preventing service-affecting failures by detecting incipient failure mechanisms or degradations and taking maintenance actions before they result in a service-affecting asset failure. Typically, a warning is issued when the asset enters a degraded state, and maintenance is performed before it fails (Figure 1.1).

In contrast, traditional systematic (scheduled) maintenance policies, whereby a maintenance action takes place after a pre-determined time, mileage or other usage variables (such as a number of operating cycles), typically lead to too much maintenance and are therefore unnecessarily high-priced in terms of both labor costs and induced downtime. Due to its significant potential for both maintenance costs reduction and system availability improvement, predictive maintenance is appealing to numerous industries, such as aerospace, automotive, rail transportation (Dersin 2018), electric power generation (including new energy sources such as wind) and electric power transmission, chemical and pharmaceutical industries, the oil and gas sector, etc. (see, e.g., Nachlas 2005).

In addition to predictive maintenance, PHM can also be relevant to asset life extension: if the RUL is known with great precision, then it can be decided, without

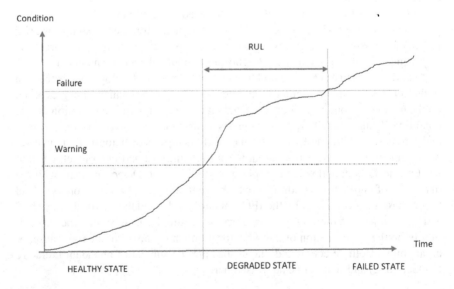

FIGURE 1.1 Use of RUL for predictive maintenance.

too high a risk, to keep an asset operating for some more time before retiring it. This question is of considerable importance to long-life assets such as rail tracks or elements of power plants for instance.

Two properties of the RUL which are of paramount importance and constitute the focus of this book are as follows:

i. its dynamic character as it generally varies over time;
ii. its uncertain character: it is best modeled as a random variable.

The fact that RUL is time-dependent is intuitively obvious: since most assets wear out with time, the time remaining to failure tends to diminish with age in general.

In addition, RUL also must be considered as a random variable, as it is affected by several sources of uncertainty:

i. epistemic uncertainty, i.e., uncertainty on the models (degradation laws; lifetime parameters);
ii. input uncertainty, including noise, sometimes called aleatoric uncertainty;
iii. uncertainty on the future load profile of the asset, and the environment it will be subjected to.

Therefore any decision based on an estimate of the RUL entails a risk, which ideally should be quantified. Risk and cost must be balanced against each other: for instance, increasing the time interval between two successive systematic preventive maintenance actions will reduce the cost of maintenance, but usually at the expense of an increased risk of failure.

If the uncertainty inherent to the RUL is not recognized, corresponding predictive maintenance strategies can be either too pessimistic (if RUL is underestimated), and therefore too costly, or too optimistic (if RUL is overestimated) and thus ineffective (as they will be less likely to prevent failures), as illustrated in Figure 1.1.

Methods to estimate RUL are very numerous and diverse. Broadly speaking, they fall into the following categories, as illustrated in Figure 1.2, borrowed from

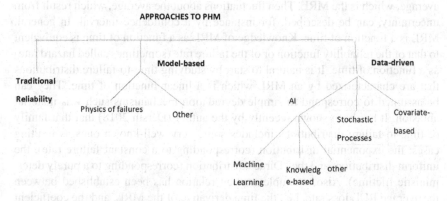

FIGURE 1.2 Various approaches to PHM.

a recent very complete survey (Jolfaei 2022): traditional reliability-based, model-based, data-driven, and finally hybrid (a combination of several approaches). Therefore, PHM is supported by reliability engineering, physics, and data science.

One can further distinguish, under model-based techniques, those which rely on physics of failures in order to model degradation processes, and other types of models. Under data-driven methods too, there are several sub-categories (from right to left in Figure 1.2): statistical covariate-based methods, such as the proportional hazard models, the best known of which is due to Cox (1972); the stochastic process approach where degradations are modeled by stochastic processes (Huynh 2017); knowledge-based approaches (expert systems), and finally machine learning (ML). The latter methods (ML) have recently become much more widely applicable thanks to the digital transformation mentioned earlier, which in turn has stimulated research on algorithms. This book will not address the entire spectrum of methods but will rather focus on reliability-based methods and stochastic processes. Our purpose indeed is not to describe extensively all PHM algorithm toolboxes; there is already an abundant and excellent literature on that subject (see e.g. Gouriveau 2016, Jolfaei 2022, Atamuradov 2017), including the latest views on potential opportunities for application of up-to-date artificial intelligence methods, such as deep learning (Fink 2020). Rather, the goal of this book is to relate PHM needs to traditional reliability engineering methods and to show how those can support PHM algorithms, sometimes in ways that had not been explored so far. Very classical results are reviewed and at the same time novel research results, including original investigations by the author, are presented.

The focus of the analysis that will be presented is the concept of RUL loss rate (Kumar 2016), namely the rate at which RUL decreases as time goes by. On average, this is measured by the time derivative of the "mean residual life" (MRL), a concept studied by reliability engineers since the 1960s (Barlow 1965) and still a fertile ground for research (Banjevic 2009, Huynh 2017). Recently, interest for this concept has been revived in the framework of survival analysis, also used in medicine (Johnson 2018). MRL typically refers to an entire population of identical assets whereas RUL is linked to one particular asset and fluctuates about the average, which is the MRL. Then fluctuations about the average, which result from uncertainty, can be described, for instance by a confidence interval. In general, MRL is a function of time. Knowledge of MRL as a function of time is equivalent to that of the reliability function or of the failure rate (sometimes called hazard rate) as a function of time. It is natural to start by studying time-to-failure distributions that are characterized by an MRL which is a linear function of time. They can be assumed to correspond to simple degradation mechanisms such was material abrasion. It has been shown recently by the author (Dersin 2018) that that family of time-to-failure distributions includes some very well-known ones as limiting cases: the exponential distribution (corresponding to a constant failure rate), the uniform distribution, and the "Dirac" distribution (corresponding to a purely deterministic lifetime). Also, a simple explicit relation has been established between the average RUL loss rate, i.e., the time derivative of the MRL, and the coefficient of variation of the RUL, for that class of family, which enables the explicit derivation of a confidence interval for the RUL. The lower the average RUL loss rate, the higher the coefficient of variation of the RUL (for instance, the exponential distribution is characterized by the highest coefficient of variation, equal to 1, and the lowest average RUL loss rate, equal to 0; and the Dirac distribution is at the opposite extreme, with a coefficient of variation of 0 and an average RUL loss rate equal to 1). This shows that, for that family of time-to-failure distributions, there is an antinomy between the time-dependence of the RUL and its stochasticity: the faster the average RUL decreases with time, the less each individual RUL fluctuates around the average.

These results are presented in Chapter 3, after a brief reminder of reliability engineering fundamentals in Chapter 2. The special case of MRL linear with time is then generalized to distributions whose MRL is piecewise-linear with time (Chapter 4). Then, a broader generalization is presented (Chapter 5): a general method is described for defining a nonlinear time transformation—a "time warping"—in such a way that, in the transformed time referential, the MRL is a linear function of time. That time transformation leaves the distribution mathematical expectation and its standard deviation unchanged. In this way, the properties derived in the special case can be transformed into properties of the RUL for most time-to-failure distributions, including well-known ones such as: the Weibull distribution, the gamma distribution, the lognormal distribution, the first-hitting time of a Wiener or a gamma degradation process; and other cases. The "metric" defined by the nonlinear time transformation is studied (Chapter 6). In most cases, the time warping corresponds to an S-curve and, among other properties, an upper bound on the average RUL loss rate is derived, for times after the inflection point. Concurrent degradation processes are then discussed (Chapter 7). Statistical methods for estimating the RUL are presented and discussed in Chapter 8, primarily the maximum likelihood estimation method. The link is made with the time transformation introduced in Chapter 5, and the methods are illustrated in examples. Maintenance policy can be optimized, or at least improved, thanks to a better understanding of RUL dynamics. In that perspective, a conceptual framework for maintenance optimization, relying on RUL, is presented in Chapter 9. Finally, advanced topics, remaining challenges and suggestions for future research are discussed in Chapter 10. The book is accessible to all readers with a college-level calculus background and an elementary knowledge of probability theory. Prior exposure to reliability theory is useful but the essential notions for our purpose are recapitulated in Chapter 2.

BIBLIOGRAPHY

Asset Management Excellence: Optimizing Equipment Life-Cycle Decisions, 2d. Ed., Edited By John D. Campbell, Andrew K.S. Jardine, & Joel McGlynn, Boca Raton, FL, Taylor & Francis, CRC Press, 2022.

Atamuradov, V., Medjaher, K., Dersin, P., Lamoureux, B., & Zerhouni, N., "Prognostics and Health Management for Maintenance Practitioners-Review, Implementation and

Tools Evaluation", *International Journal of Prognostics and Health Management 8(Special Issue on Railways & Mass Transportation)*, 31, December 2017.

Banjevic, D., "Remaining Useful Life in Theory and Practice", *Metrika*, 69(2), 337–349. 2009.

Barlow, R.E., & Proschan F., *Mathematical Theory of Reliability*. New York, Wiley, 1965.

Cox, D.R., "Regression Models and Life-Tables (with Discussion)", *Journal of Royal Statistical Society Series B*, 34, 187–202, 1972.

Dersin, P., "The Class of Life Time Distributions with a Mean Residual Life Linear in Time: Application to PHM", in *Safety & Reliability: Safe Societies in a Changing World*, Eds S. Haugen, A. Barros et al., London, CRC Press, 1093–1099, 2018.

Dersin, P. et al., "Prognostics & Health Management in Railways", Ch. 6, in *Handbook of RAMS in Railway Systems*, Eds K. Mahboob & E. Zio, Boca Raton, CRC Press, Taylor & Francis, 79–98, 2018.

Fink, O., Wang, Q., Svensén, M., Dersin, P., Lee, W-J., & Ducoffe, M., "Potential, Challenges and Future Directions for Deep Learning in Prognostic and Health Management Applications", in *Engineering Applications of Artificial Intelligence*, 92, 2020, 103678, ISSN 0952-1976.

Finkelstein, M., *Failure Rate Modeling for Reliability and Risk*. Switzerland, Springer, 2008.

Forest, F., "Unsupervised Learning of Data Representations and Cluster Structures: Applications to Large-Scale Health Monitoring of Turbofan Aircraft Engine", Ph.D. Thesis, Université Sorbonne Paris Nord, 2021.

Gouriveau, R. et al., *From Prognostics and Health Management to Predictive Maintenance*. Hoboken, NJ, J. Wiley & Sons, 2016.

Huynh, K.T, Gralle, A., & Bérenguer, C. "Assessment of Diagnostic and Prognostic Condition Indices for Efficient and Robust Maintenance Decision-Making of Systems Subject to Stress Corrosion Cracking", in *Reliability Engineering & System Safety*, 159, 237–254, 2017.

Huynh, K., Grall, A., & Bérenguer, C. "A Parametric Predictive Maintenance Decision-Making Framework Considering Improved System Health Prognosis Precision", in *IEEE Transactions on Reliability*, 68(1), 375–396, 2019. 10.1109/TR.2018.2829771. hal-01887.

IEC 2015: International Electrotechnical Commission, IEC-60060-192, "International electrotechnical vocabulary-Part 192: dependability", 2015.

ISO 55000:2014, Asset Management — Overview, principles and terminology, 2014.

Johnson, L.L, "Introduction to Survival Analysis", Ch. 26, in *Principles and Practice of Clinical Research*, Gallin J.L et al., London, Academic Press, 373–381, 2018.

Jolfaei, N., Rameezdeen, R., Gorjian, N., Bo Jin, B. & Chow, C. "Prognostic Modelling for Industrial Asset Health Management", *Safety & Reliability*, 41(1), 45–97, 2022.

Kumar, U., et al. *Current Trends in Reliability, Availability, Maintainability and Safety—An Industry Perspective*. Switzerland, Springer International Publishing, 2016.

Lee, J. et al., "Prognostics and Health Management Design for Rotary Machinery Systems—Reviews, Methodology and Applications", *Mechanical Systems and Signal Processing*, 42(1–2), 314–334, 2014.

Nachlas, J., *Reliability Engineering: Probabilistic Models and Maintenance Methods*. Boca Raton, FL, CRC Press, Taylor & Francis, 2005.

Saxena, A., et al., "Metrics for Evaluating Performance of Prognostic Techniques", *2008 International Conference on Prognostics and Health Management*, 2008, pp. 1–17.

Si, X., et al., *Data Driven Remaining Useful Life Prognosis Techniques-Stochastic Models, Methods & Applications*. Berlin, Heidelberg, Springer, 2017.

Zio, E., "Prognostics and Health Management (PHM): Where Are We and Where Do We (need to) Go in Theory and Practice", *Reliability Engineering & System Safety*, 218, Part A, 2022. ISSN 0951-8320.

2 Reminder of Reliability Engineering Fundamentals

2.1 RELIABILITY, FAILURE RATE, RUL, MRL, AND RUL LOSS RATE

For clarity, a number of fundamental reliability notions are recapitulated.

For more details, the reader is referred to many excellent books on the topic, listed at the end of this chapter.

2.1.1 RELIABILITY, FAILURE, AND FAILURE RATE

The concept of reliability is linked to the notion of function.

The International Electrotechnical Committee (IEC 60050-192) defines the concepts of reliability and failure as follows:

> **Definition 2.1.** Reliability is the ability to perform as required, without failure, for a given time interval, under given conditions.
> **Definition 2.2.** A failure is the loss of the ability to perform as required.

Thus a failure need not be a sudden catastrophic event, a so-called hard failure (something that breaks down): the occurrence of a failure is synonymous with the loss of function, i.e. the fact that an item of equipment, or a system, no longer performs its function according to specifications, i.e. "as required".

Let us take a few simple examples to illustrate that concept.

A heating, ventilation and air conditioning unit must deliver a flow of clean air measured by its throughput, in m^3/h. When this throughput diminishes, due to various factors such as the clogging of air filters with dust, there comes a time when the desired throughput is no longer delivered, and therefore the cooling function is no longer assured.

Thus the failure occurs, by definition, when the air throughput falls below a threshold which is defined by the technical specifications of the cooling unit.

Another example (studied in Chapter 8) is that of a light-emitting diode (LED). The LED emits light, which is measured in lumen. Due to various stresses such as temperature and current, the luminous flux will decrease over time. Once that flux

DOI: 10.1201/9781003250685-2

decreases under a given threshold (which corresponds to the specified flux), the equipment no longer performs as required (even though it is still emitting light), and therefore it is considered to have failed.

Still another example is provided by a brake pad in railways. Brake pads do not perform their function efficiently once, due to wear, their thickness diminishes beyond a certain threshold. Thus in that case the failure occurs when the brake pad thickness crosses that threshold.

One may also think of cracks in various types of infrastructure, for instance rails, or protective containers in power plants: beyond a certain crack size, the rail track no longer ensures safely the adequate contact with train wheels, or the container will let fluid (such as gas) escape. In both cases, they no longer perform as required.

Predictive maintenance clearly can be considered for systems that degrade progressively rather than for those which undergo sudden failures.

Let us notice two important qualifiers in Definition 2.1: "for a given time interval" and "under given conditions". Both are important.

First then, the concept of reliability is intimately tied to that of time. It is meaningless to speak about the reliability of an item without specifying the period of time over which that reliability is considered. (Alternatively, instead of time, an other usage variable can be utilized, such as mileage or number of cycles.)

Second, "given conditions" include aspects that affect reliability, such as: mode of operation, stress levels, environmental conditions, and maintenance. For instance, the LED operation will be affected by stresses such as current and temperature. A train access door which is opened and closed 1000 times a day will not last as long, without failure, as one which only undergoes ten opening–closing cycles a day. An aircraft engine reliability will be affected by the number of "hard landings" which they undergo, and by the type of trips which they make and the weather they frequently encounter. And finally, it is clear that maintenance, or the lack of it, generally impacts reliability (maintenance will be addressed in Chapter 9).

Therefore all of those conditions must be taken into account when predicting reliability or estimating it from field data.

So far, the definition given has been rather qualitative. It can be made quantitative by introducing probabilities.

Definition 2.3
The reliability function of an item at time t, $R(t)$, is defined as the probability that that item, assumed to be operational at time 0, operates without failure until at least time t.

Or, if T denotes the random variable "time to failure" (TTF)

$$R(t) = P[T > t]. \tag{2.1}$$

As just mentioned, reliability is impacted by external conditions and stresses. Mathematically, those can be described by variables sometimes called "covariates".

$$R(t) = R(t, S_1, S_2, \ldots S_m) \tag{2.2}$$

for m covariates $S_1, S_2, \ldots S_m$ (which themselves can be time-dependent).

Likewise, the concept of failure can be quantified by the failure rate (sometimes called hazard rate).

The notion of failure rate is similar to that of mortality rate used by insurance companies.

It arises by asking the question: "given that this item of equipment has been operating according to specifications for so many days, what is the probability that it fails ('dies') tomorrow?"

Mathematically, the notion is captured through a conditional probability, or more precisely, a conditional probability per unit of time (or of usage variable).

Formally, the definition is the following (also to be found in IEC 2015 for instance).

Definition 2.4

The failure rate is the limit, if it exists, of the quotient of the conditional probability that the failure of a non-repairable item occurs within time interval $(t, t + \Delta t)$ by Δt, when Δt tends to zero, given that failure has not occurred within time interval $(0, t)$.

The failure rate at time t is traditionally denoted $\lambda(t)$.

There follows from Definition 2.4 the link between failure rate and reliability function:

$$\lambda(t) = -\frac{\dfrac{dR}{dt}}{R(t)}. \tag{2.3}$$

Indeed from Definition 2.4

$$P[\text{failure in } (t, t + \Delta t) \text{ |no failure in } (0, t)] = P[t < T < t + \Delta t \,|\, T > t] = \frac{R(t) - R(t + \Delta t)}{R(t)}$$

and (2.3) is obtained by taking the limit for $\Delta t \to 0$.

The ratio $-\dfrac{dR}{dt}$ is the probability density of the TTF, usually denoted $f(t)$.

As $R(t)$ is a non-increasing function of time, $\dfrac{dR}{dt} \leq 0$.

Thus

$$-\frac{dR}{dt} = f(t) \geq 0 \tag{2.4}$$

and

$$\lambda(t) = \frac{f(t)}{R(t)}. \tag{2.5}$$

For $0 \le a < b$,

$$\int_a^b f(t)\,dt = R(a) - R(b). \qquad (2.6)$$

In particular,

$$\int_0^\infty f(t)\,dt = R(0) - R(\infty) = 1. \qquad (2.7)$$

Note that the definition refers to "non-repairable items". This means that the failure rate refers to the first failure. If the item is repairable, and therefore typically undergoes several failures over time, its reliability is characterized by a *failure intensity*, which is the average number of failures per unit of time.

In this book, up to and including Chapter 8, non-repairable items will be considered.

2.1.2   RUL, MRL, MTTF, AND RUL LOSS RATE

The concept of remaining useful life (RUL) is fundamental for the elaboration of any predictive maintenance policy. Essentially, given that an asset has been operating for some time without failure, its RUL is the time during which it will keep operating without failure, or equivalently, the time to failure from then on.

The RUL is a random variable and a function of time.

It is a function of time because, in general, how long the asset will keep operating without failure is a function of its age: most items of equipment tend to wear out with time: a machine which has already been in operation for 40 years is usually not likely to keep operating failure-free for another 40 years. Thus RUL is usually a decreasing (at least non-increasing) function of time; although, in some circumstances, it can be decreasing (for instance if the asset is in a less stressful environment than previously). And RUL is a random variable because failures are generally the outcomes of phenomena that are best described as random processes: degradations subject to various stresses, as well as variations in mission profile.

The mean residual life (MRL) is then the mathematical expectation of the RUL, i.e. the expected remaining life of an asset that has been operating for some time without failure.

It is a number, function of time. Unlike the RUL, it is not a random variable, it is the expectation of a random variable.

One can see the MRL as referring to a population of identical assets (for instance all engines of a fleet of aircraft), while each individual asset (each particular engine) has its own RUL.

Thus MRL(t) defines the average aging trend in a population consisting of many identical assets, while the RUL (t) function of a given asset describes the aging of that asset only.

There can be sometimes wide fluctuations across the various assets, around the mean (MRL) of the population: just as in a human population, some can live much longer than average for various reasons (better resistance to stress, life style, etc.).

In traditional reliability engineering, one could only deal with populations. But with the advent of cost-effective wireless sensors and communication, the Internet of Things, it is now possible to monitor assets individually, hence the interest in making individual prediction and performing "bespoke maintenance": therefore RUL, rather than just MRL, becomes important.

One can write

$$\mathrm{MRL}(t) = E\big(\mathrm{RUL}(t)\big), \tag{2.8}$$

where MRL (t) is the equivalent to what actuaries call the life expectation at age t.

For $t = 0$, MRL is equal to the mean time to failure, MTTF.

$$\mathrm{MRL}(0) = \mathrm{MTTF}. \tag{2.9}$$

This notion is analogous to what actuaries call the life expectation at birth.

Actually, MTTF now is increasingly meant to stand for mean *operating* time to failure (IEC 2015), to emphasize the fact that what matters in aging is operating time, not calendar time. Also, time can be replaced with alternative meaningful usage variables, such as mileage or number of cycles, depending on context. (For instance, for train wheels, mileage is the most appropriate usage variable.) The same remark applies to MRL.

From now on, for convenience, the notation $V(t)$ will be used instead of MRL(t).

The properties of MRL have been studied since the very first days of reliability engineering (Barlow & Proschan 1965).

As RUL is a function of time, it is useful to consider its derivative $\dfrac{d\mathrm{RUL}}{dt}$, which will be called the *RUL loss rate*: it measures the rate at which RUL changes (usually decreases) as time goes by.

Knowing the RUL loss rate is useful for the decision maker as it is an indication of how the risk of failure evolves with time. It has immediate implications for maintenance and, more generally, asset management and risk management. For instance, in the infrastructure crack growth example mentioned earlier, an increase in RUL loss rate will be highly correlated with an acceleration of crack growth.

The derivative of the MRL is equivalent to the expected value of the RUL loss rate. In fact, by interchanging the derivatives and expectation operators the following relationship can be derived:

$$E\left(\frac{d\mathrm{RUL}}{dt}\right) = \frac{dE(\mathrm{RUL})}{dt} = \frac{d\mathrm{MRL}}{dt} \equiv \frac{dV}{dt}. \tag{2.10}$$

Thus the time derivative of the MRL will play an important role in the study of the RUL dynamics, i.e. of how RUL is expected to evolve over time.

Now the time derivative of the MRL, the reliability function $R(t)$ and the failure rate $\lambda(t)$ are related by a simple relation, as shown in the following section.

2.2 FUNDAMENTAL RELATION BETWEEN MRL, RELIABILITY FUNCTION AND FAILURE RATE

Theorem 2.1

If $V(t)$ denotes the MRL at time t, $R(t)$ the reliability function at time t, and $\lambda(t)$ the failure rate at time t, the following relation holds:

$$\frac{dV}{dt} = \lambda(t)V(t) - 1 \tag{2.11}$$

Proof

From the definitions, there follows that

$$V(t) = \frac{\int_t^\infty R(u)\,du}{R(t)} \tag{2.12}$$

and, in particular, for $t = 0$, since $R(0) = 1$,

$$V(0) = \text{MTTF} = \int_0^\infty R(u)\,du. \tag{2.13}$$

On the hand, by definition, the failure rate $\lambda(t)$ is related to the reliability function by (2.3)

From (2.12), using the differentiation rules and (2.3) and denoting $R'(t)$ the derivative of $R(t)$

$$\frac{dV}{dt} = \frac{-R^2(t) - R'(t)\int_t^\infty R(u)\,du}{R^2(t)} = -1 + \lambda(t)V(t) \quad \text{qed.}$$

Any one of the three functions $R(t)$, $\lambda(t)$, and $V(t)$ uniquely determines the other two.

From $R(t)$, $\lambda(t)$ is derived by means of (2.3), and $V(t)$ by means of (2.12).

From $\lambda(t)$, $V(t)$ is derived by solving the linear differential (2.11) and $R(t)$ by inverting (2.3), which yields

$$R(t) = e^{-\int_0^t \lambda(u)\,du} \tag{2.14}$$

Sometimes, the exponent is called the cumulative hazard function, denoted $\Lambda(t)$:

$$\Lambda(t) = \int_0^t \lambda(u). \tag{2.15}$$

(Sometimes $\lambda(t)$ is called hazard rate instead of failure rate.)

Then (2.14) can be written:

$$R(t) = e^{-\Lambda(t)}. \tag{2.16}$$

From $V(t)$, the failure rate $\lambda(t)$ is obtained directly by means of (2.11):

$$\lambda(t) = \frac{1 + V'(t)}{V(t)}. \tag{2.17}$$

And finally, the reliability function $R(t)$ can also be derived directly from the MRL, i.e. $V(t)$:

$$R(t) = \frac{V(0)}{V(t)} e^{-\int_0^t \frac{dx}{V(x)}}. \tag{2.18}$$

(Swartz 1973)

The three functions describe the same reality of asset life from different viewpoints.

The reliability function answers the question, for any date in the future, how likely it is that the item will still be operational on that date. The failure rate measures the probability for an item to fail, at a given instant, per unit of time, given that it has not failed yet. And the MRL is simply the expectation of the remaining life. Of the three measures, it is perhaps the easiest to comprehend intuitively, although the failure rate is used much more often.

To pursue the analogy with actuarial science, the failure rate is equivalent to a mortality rate.

Interpretation of Theorem 2.1

Equation (2.11) can be written as follows in differential form:

$$dV = -dt + \lambda(t)V(t)dt = -(1 - \lambda(t)V(t))dt. \tag{2.19}$$

The left-hand side is the expected variation in RUL over a small time interval dt.

Usually, dV is negative. Therefore the right-hand side shows that the expected loss of RUL over time dt is usually less than dt.

In other words, over a span of say, one day, the asset remaining life decreases by less than one day in general. How much less is quantified by (2.19) and depends on the reliability characteristic of the asset.

Illustrations

Two examples are now considered to illustrate the above results.

1. The first one is a degenerate case: the case when the failure rate remains constant with time:

$$\lambda(t) = \lambda. \tag{2.20}$$

Then the reliability function, from (2.14), is given by

$$R(t) = e^{-\lambda t} \quad \text{for} \quad t \geq 0. \tag{2.21}$$

Thus the time to failure is exponentially distributed.

From (2.12), it follows immediately that the MRL is constant too, and

$$\text{MRL}(t) \equiv V(t) = \text{MTTF} \quad \text{for all values of } t \tag{2.22}$$

From (2.11) or (2.12) then:

$$V(t) = \text{MTTF} = \frac{1}{\lambda}. \tag{2.23}$$

Conversely, from (2.11) it can be shown that, if the MRL is constant in time, then the failure rate is constant and the time to failure is exponentially distributed. This follows immediately from the fact that, if $V(t)$ is constant, then $\dfrac{dV}{dt} = 0$ and, from the right-hand side of (2.11), λ is the reciprocal of V. Also, it is verified that application of (2.16) then yields the exponential function for $R(t)$.

Thus the exponential distribution for the time to failure is characterized by a constant MRL and by a constant failure rate.

2. Consider now the case when the failure rate increases linearly with time:

$$\lambda(t) = \alpha t \quad (\text{with } a > 0). \tag{2.24}$$

This corresponds to the Rayleigh distribution for the time to failure, a special case of the Weibull distribution for which the shape factor β is equal to 2 (see e.g. Birolini 2017).

Then (2.11) becomes:

$$\frac{dV}{dt} = \alpha t V - 1. \tag{2.25}$$

It is convenient to express a in terms of the scale parameter η of the Weibull distribution:

$$a = \frac{2}{\eta^2}. \tag{2.26}$$

Then the MTTF is obtained:

$$\text{MTTF} = \eta\Gamma\left(1+\frac{1}{\beta}\right),$$

where $\Gamma(.)$ denotes Euler's gamma function. For $\beta = 2$, $\Gamma\left(\frac{3}{2}\right) = \frac{\sqrt{\pi}}{2}$
and

$$V(0) = \text{MTTF} = \eta\frac{\sqrt{\pi}}{2}. \tag{2.27}$$

Solving the linear differential (2.11) with initial condition (2.27) results in

$$V(t) = \eta\sqrt{\pi}\left[1 - \Phi\left(\frac{\sqrt{2}}{\eta}t\right)\right]e^{\left(\frac{t}{\eta}\right)^2}, \tag{2.28}$$

where $\Phi(.)$ denotes the cumulative distribution function of the standard normal deviate:

$$\Phi(x) = \frac{1}{\sqrt{(2\pi)}}\int_{-\infty}^{x}e^{-\frac{s^2}{2}}\,ds. \tag{2.29}$$

A plot of $V(t)$ as a function of time is shown in Figure 2.1. $V(t)$ is seen to be a decreasing convex function of time.

For instance, after a time equal to MTTF, it is found that

$$V(MTTF) = 2V(0)\left[1 - \Phi\left(\frac{\sqrt{\pi}}{2}\right)\right]e^{\frac{\pi}{4}} = 0.46V(0).$$

In general, no closed form solution for MRL(t) can be obtained: numerical solutions and simulations must be used.

2.3 CONFIDENCE INTERVAL FOR RUL. ILLUSTRATIONS ON SPECIAL CASES

Confidence intervals can be derived from the probability distribution of the RUL.

The probability distribution of the RUL is obtained from the following relationship, which results directly from the definitions:

$$P\left[\text{RUL}(t) > s\right] = P\left[T > t + s|T > t\right] = \frac{P[T > t+s]}{P[T > t]} = \frac{R(t+s)}{R(t)}. \tag{2.30}$$

Hence the probability density function of the RUL is obtained by differentiation:

$$f_{\text{RUL}(t)}(s) = -\frac{dP\left[\text{RUL}(t) > s\right]}{ds} = \frac{f(t+s)}{R(t)}. \tag{2.31}$$

Confidence intervals can thus be derived.

A $(1-\delta)$-level confidence interval for RUL(t) can be derived as the interval (a, b) such that

$$P\left[\text{RUL}(t) < a\right] = \frac{\delta}{2} \tag{2.32a}$$

and

$$P\left[\text{RUL}(t) > b\right] = \frac{\delta}{2}. \tag{2.32b}$$

Let us apply this to the two examples of Section 2.2.

For the exponential distribution,

$$P\left[\text{RUL}(t) > s\right] = \frac{R(t+s)}{R(t)} = e^{-\lambda s}, \tag{2.33}$$

which shows that the exponential function has no memory: the RUL is distributed just as the initial survival time.

The bounds of the $(1 - \delta)$ confidence interval are obtained straightforwardly from (2.30):

$$P\left[\text{RUL}(t) < a\right] = \frac{\delta}{2}$$

$$P\left[\text{RUL}(t) > b\right] = \frac{\delta}{2}$$

hence

$$a = -\frac{\ln\left(1 - \frac{\delta}{2}\right)}{\lambda} \tag{2.34}$$

$$b = -\frac{\ln\left(\frac{\delta}{2}\right)}{\lambda}. \tag{2.35}$$

The confidence interval is independent of t, which is a well-known characteristic of the exponential distribution.

In the case of the Rayleigh distribution, confidence bounds can also be derived from the reliability function but, as seen below, the situation is quite different:

$$P\left[\text{RUL}(t) > s\right] = \frac{R(t+s)}{R(t)} = \frac{e^{-\left(\frac{t+s}{\eta}\right)^2}}{e^{-\left(\frac{t}{s}\right)^2}}. \tag{2.36}$$

The bounds a and b of the confidence interval are then given, in the same way as before, by setting

$$P\left[\text{RUL}(t) > b\right] = \frac{\delta}{2}$$

and

$$P\left[\text{RUL}(t) < a\right] = 1 - \frac{\delta}{2}.$$

There follows:

$$a = \left[t^2 - \eta^2 \ln\left(1 - \frac{\delta}{2}\right)\right]^{\frac{1}{2}} - t \tag{2.37}$$

$$b = \left[t^2 - \eta^2 \ln\left(\frac{\delta}{2}\right)\right]^{\frac{1}{2}} - t. \tag{2.38}$$

In contrast with the exponential distribution, the confidence interval for the RUL for a Rayleigh-distributed time to failure is a function of time.

Its width decreases as t increases.

Figure 2.1 shows the plot of MRL as a function of time as well as the 80% confidence interval for the RUL.

For instance, corresponding to t= 130h, $V(t)$ =31.7h but the 80% confidence interval is defined by (4h, 69.8h), thus there is an 80% chance that 4h < RUL< 69.8h, with an expectation (MRL) of 31.7h.

As time increases, the size of the confidence interval shrinks (and correspondingly the MRL decreases, eventually to 0).

For instance, for t = 300h, MRL=15.8h and the 80% confidence interval is given by (1.8h, 37h).

Remark on Confidence Interval

In the above calculation, the only information available on the asset is that its time to failure has a Rayleigh distribution and it has not failed yet. When condition data are acquired (in real time or not), the RUL prediction can be made more precise, which results in a narrower confidence interval. But it is important to remember

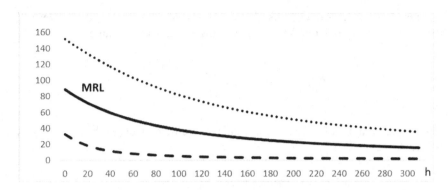

FIGURE 2.1   MRL and 80% confidence interval (a, b) for RUL, versus time (Rayleigh distribution).

that uncertainty can never be eliminated, therefore there is always a confidence interval associated with any RUL estimate.

Remark on the Definition of Failure
Let us recall again that, from the definition of the concept of reliability, a failure is a loss of function. In some cases, the condition of an asset is assessed by a health indicator and the relevant function is lost when the health indicator crosses a pre-defined threshold. Then, the time to failure is the time needed for the indicator to cross the threshold, or "first hitting time". The probability distribution of the first hitting time is then the key for determination of RUL and MRL.

2.4   EXERCISES

1. Consider the exponential time-to-failure distribution, with parameter $\lambda = 10^{-3}$/h.
   Calculate the probability of survival (without failure) beyond 100h and beyond 1000h.
   Calculate the MTTF.

2. Consider a uniform distribution on the time interval (a, b) with $0 \le a < b$, defined by the following probability density function:

$$f(t) = \begin{cases} \dfrac{1}{b-a} & a \le x \le b \\ 0 & x < a \, or \, x > b \end{cases}.$$

   Assume the time to failure of an asset is modeled by that distribution.
   a. Express the reliability function, $R(t)$, in terms of a and b.
   b. Express the failure rate, $\lambda(t)$, in terms of a and b.
   c. Express the MRL function, $V(t)$, in terms of a and b.

    d. Let $a = 0$ and $b = 2000$h. Calculate the probability that the asset survives without failure up to at least 100h; up to at least 1000h. Calculate the MTTF and the value of MRL(t) for $t= 1000$h.

3. Sometimes the B_{10} value, or 10% quantile, is used. It is by definition the time which is such that the probability of a failure occurring before that time is 10% .

    a. Calculate B_{10} for the exponential distribution of Exercise 1.

    b. Calculate B_{10} for the uniform distribution of Exercise 2 (with $a = 0$ and $b = 2000$h).

4. As a reminder, the reliability function of the Weibull distribution with scale parameter η and shape parameter β is given by

$$R(t) = e^{-\left(\frac{t}{\eta}\right)^{\beta}}.$$

Calculate B_{10} for the Weibull distribution with $\eta=1000$h and $\beta=3$.

5. The B_{50} value, or median, is defined by the property that the probability of the random variable exceeding that value is ½ (in other words it divides the population into two equal halves).

Calculate the median for:

    a. The exponential distribution of Exercise 1.

    b. The uniform distribution of Exercise 2.

    c. The Weibull distribution of Exercise 3.

    d. The Rayleigh distribution with scale factor $\eta=100$h.

    e. In each case, compare the median with the mean; for which distributions are those two values equal?

6. For the four distributions of Exercise 5,

    a. calculate the probability of surviving beyond time $t=$MTTF;

    b. calculate the MRL at $t=$ MTTF;

    c. draw conclusions about the suitability of MTTF as a reliability risk measure.

BIBLIOGRAPHY

Barlow, R.E., & Proschan, F., *Mathematical Theory of Reliability*. New York, Wiley 1965.

Birolini, A., *Reliability Engineering: Theory and Practice*. Berlin, Springer, 8th Edition, 2017.

IEC 60050-192, International Electrotechnical Vocabulary – Part 192: Dependability, 2015.

Nachlas, J.A., *Reliability Engineering – Probabilistic Models & Maintenance Methods*. Boca Raton, FL, Taylor & Francis, CRC Press, 2005.

Swartz, G.B., "The Mean Residual Lifetime Function", *IEEE Transactions on Reliability*, 22(2), 108–109, 1973.

3 The RUL Loss Rate for a Special Class of Time-to-Failure Distributions*

MRL Linear Function of Time

3.1 CHARACTERIZING THE SPECIAL FAMILY OF DISTRIBUTIONS

We now study the family of time-to-failure (TTF) distributions characterized by the property that their mean residual life $MRL(t)$ is a linear non-increasing function of time (or whatever other relevant usage variable, such as mileage).

A priori, the interest in this family of distributions can be justified by the consideration of simple degradation processes, such as carbon strip wear, wheel wear, and so on.

Using the notation introduced in Chapter 2, such a family is characterized by the following equation:

$$\frac{dV}{dt} = -k \qquad (3.1)$$

with $0 \le k \le 1$.

This family is of interest, first of all because linearity characterizes simple degradation phenomena; and second—as will be seen below—because it includes among its members some extremely well-known probability distributions, such as the exponential distribution and the uniform distribution.

The parameter k is then the average remaining useful life (RUL) loss rate—assumed to be constant in this special case. Thus, one can write, for this special family:

$$dV = -k\,dt$$

The average loss of RUL in a time increment dt is a fraction k of that time increment.

For instance, in the crack growth example, with the TTF proportional to the crack size, k would measure the rate of growth of the crack. More generally, k can be seen somehow as a degradation speed.

For convenience, from now on, the MTTF will be denoted by the symbol μ. Thus,

---

* Parts of Sections 3.1 to 3.5 reproduced from Dersin (2018), with kind permission from CRC Press.

DOI: 10.1201/9781003250685-3

$$V(0) = \text{MTTF} = \mu. \tag{3.2}$$

From (3.1) and (3.2), it is straightforward that

$$V(t) = \mu - kt. \tag{3.3}$$

Necessarily, $V(t) \geq 0$ since $V(t)$ is the expectation of a duration.

Therefore, for assets with linearly decreasing MRL, the support of the time variable is the interval $[0, \frac{\mu}{k}]$ since a value greater than $\frac{\mu}{k}$ would make $V(t)$ negative.

As time t increases from 0 to $\frac{\mu}{k}$, the MRL decreases from μ to 0:

$$0 \leq V(t) \leq \mu. \tag{3.4}$$

Thus the special family of distributions under study in this chapter is characterized by a finite support. Note that the exponential distribution for the TTF is part of the family, corresponding to $k = 0$. For that case, the support of the time variable coincides with the nonnegative real half line.

From Theorem 2.1 of Chapter 2 (2.17), and from (3.3), there follows that the failure rate for a distribution belonging to the special family is given by

$$\lambda(k;t) = \frac{1 + \dfrac{dV}{dt}}{V(t)} = \frac{1-k}{\mu - kt} \quad \text{for} \quad 0 \leq t \leq \frac{\mu}{k}. \tag{3.5}$$

From the general relation between failure rate and reliability function (2.14), the reliability function is derived:

$$R(k;t) = e^{-\int_0^t \frac{1-k}{\mu-ks} ds} \quad \text{for} \quad 0 \leq t < \frac{\mu}{k} \tag{3.6}$$

and

$$R(k;t) = 0 \quad \text{for} \quad t \geq \frac{\mu}{k}.$$

From

$$\frac{ds}{\mu - ks} = d\left[\frac{\log(\mu - ks)}{-k}\right],$$

there follows:

$$R(k;t) = e^{\frac{(1-k)\left[\log\mu - \log(\mu - kt)\right]}{(-k)}}$$

$$= e^{\left(1-\frac{1}{k}\right)\log\left(\frac{\mu}{\mu - kt}\right)}$$

which can be expressed as

$$R(k;t) = \begin{cases} \left(\dfrac{\mu}{\mu - kt}\right)^{1-\frac{1}{k}} & \text{for } 0 \le t < \dfrac{\mu}{k}. \\ 0 & \text{for } t > \dfrac{\mu}{k} \end{cases} \quad (3.7)$$

Thus the two-parameter family of probability distributions (parameters μ and k) is fully characterized by (3.3), (3.5), or (3.7).

3.2 LIMITING CASES

Three special cases are now examined, corresponding respectively to the values 0, 1, and ½ for the parameter k. In other words, we shall be studying in this section systems which, on average, lose 0, one hour or half an hour of RUL, per hour of service.

Conceivably, one might also consider cases when $k > 1$, i.e. systems which on average lose more than one hour of remaining life for each hour of service; but, from (3.5), that would make sense only on the range $t > \mu / k$, otherwise the failure rate would be negative. Also, with $k < 0$, the model would apply to a reliability growth context, i.e. MRL(t) increasing with t. In Prognostics and Health Management (PHM), one is rather more interested in degradations and therefore an MRL function which decreases with time.

Those cases will therefore not be studied in this book.

3.2.1 EXPONENTIAL DISTRIBUTION

If the parameter k is set to 0, the support of the variable t becomes the half real line $(0, \infty)$.

The following limit is obtained from (3.5):

$$\lambda(0;t) = \frac{1}{\mu}. \quad (3.8)$$

It can also be verified from (3.7) that:

$$\lim_{k \to 0} R(k,t) = e^{-\frac{t}{\mu}}.$$ (3.9)

Those properties characterize the exponential distribution.

As seen previously and confirmed from (3.3), the MRL is then constant in time: $V(t) = \mu = $ MTTF. The limiting character of the exponential distribution has been studied in Chapter 2 already.

3.2.2 DIRAC DISTRIBUTION

At the other extreme, if $k = 1$, the evolution of the MRL over time is characterized by

$$V(t) = \mu - t \quad \text{for} \quad 0 \le t < \mu$$ (3.10)

$$0 \qquad \text{for} \quad t \ge \mu$$

or

$$V(t) = \text{Max}(\mu - t; 0)$$ (3.11)

which means that, over any time interval Δt, the loss of MRL is precisely equal to Δt. The time parameter ranges from 0 to μ.

Equation (3.5) shows that the failure rate remains equal to zero as long as $t < \mu$ and jumps to infinity as t approaches the life limit μ.
More precisely,

$$\lim_{k \to 1} \left[\lim_{t \to \frac{\mu}{k}} \lambda(k;t) \right] = +\infty$$ (3.12)

as can be seen by setting

$$k = 1 - \varepsilon$$

$$t = \frac{\mu}{k} - \delta$$

and letting first δ, then ε, go to zero.

Likewise the reliability function has a discontinuity at $t = \mu$, where it jumps from 1 to 0, which corresponds to a *deterministic lifetime* equal to μ. It can also be verified that the right-hand side of (2.11) (Chapter 2) is always equal to -1, meaning that, over any time interval Δt, the loss of MRL is precisely equal to Δt.

3.2.3 UNIFORM DISTRIBUTION

The intermediate case $k = \dfrac{1}{2}$ defines the uniform distribution, over the range $(0, 2\mu)$.

Indeed, from (3.7), the corresponding reliability function is given by

$$R\left(\frac{1}{2};t\right) = \text{Max}\left(1 - \frac{t}{2\mu};0\right).$$ (3.13)

The MRL (3.3) is given by

$$V(t) = \text{Max}\left(\mu - \frac{t}{2};0\right)$$ (3.14)

which states that, over a time interval Δt, the loss of MRL is equal to $\frac{\Delta t}{2}$. The failure rate is given, as a function of time, by

$$\lambda\left(\frac{1}{2};t\right) = \frac{1}{2\mu - t}.$$ (3.15)

for $t \le 2\mu$.

3.3   SYNOPSIS

The characteristics of the two-parameter family of TTF distributions with an MRL linear function of time are summarized in Table 3.1 and Figure 3.1.

Thus, although all those lifetime distributions have the same MTTF, i.e., the same expectation, they differ widely in terms of failure rate evolution or expected RUL loss rate, i.e. time derivative of their MRL: at one end of the spectrum (the exponential distribution, $k = 0$), there is no expected loss of RUL, i.e. the MRL remains constant with time; at the other end of the spectrum (the Dirac distribution, $k = 1$), every passing second marks a one-second average loss in RUL. Half way,

---

TABLE 3.1
Times to Failure with MRL Linear Function of Time

Value of k	Range of t	$R(k; t)$	$V(t)$	$\lambda(k;t)$	dV/dt	MTTF
$0 \le k \le 1$	$[0, \frac{\mu}{k}]$	$\left(\frac{\mu}{\mu - kt}\right)^{1 - \frac{1}{k}}$	$\mu - kt$	$\frac{1-k}{\mu - kt}$	$-k$	μ
0 (exponential)	$[0,\infty)$	$e^{-\frac{t}{\mu}}$	μ	$\frac{1}{\mu}$	0	μ
½ (uniform)	$[0, 2\mu]$	$1 - \frac{t}{2\mu}$	$\mu - \frac{t}{2}$	$\frac{1}{2\mu - t}$	$-\frac{1}{2}$	μ
1 (Dirac)	$[0, \mu]$	$1\ t<\mu$ $0\ t\ge\mu$	$\mu - t$	$0\ t<\mu$ $\infty\ t = \mu$	-1	μ

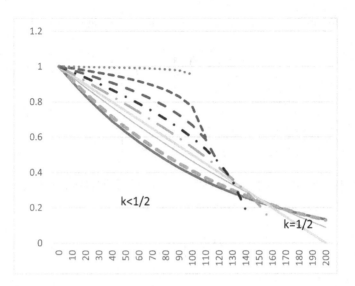

FIGURE 3.1   Reliability function for various values of *k*.

with the uniform distribution (*k* = ½), every passing second results in a half-second loss in RUL on average.

It is to be noticed (Figure 3.1 and equation 3.7) that, for *k* < 1/2, the reliability function is strictly convex; for *k* > 1/2, it is strictly concave; and, for *k*=1/2 (the uniform distribution), it is linear. Thus the uniform distribution is a sort of borderline case within the family of distributions with MRL linear in time: it separates them between fast degradation dynamics (*k* > 1/2) and slow degradation dynamics (*k* < 1/2).

Since the considered family has just 2 parameters (*μ* and *k*), another way of characterizing it is through its first- and second-order moments. The first-order moment is *μ*. It will now be seen that the coefficient of variation, i.e. the ratio of standard deviation to mean, is uniquely determined by *k*.

3.4   RUL DISTRIBUTION, COEFFICIENT OF VARIATION OF TTF AND RUL

First, we shall be studying the coefficient of variation of the TTF; and subsequently that of the RUL.

3.4.1   Coefficient of Variation of the Time to Failure

The probability density function *f*(*k*; *t*) for the TTF is obtained from (3.5) and (3.7):

$$f(k;t) = \lambda(k;t)R(k;t) \tag{3.16}$$

$$f(k;t) = \frac{1-k}{\mu - kt} \left(\frac{\mu}{\mu - kt} \right)^{1-\frac{1}{k}}.$$ (3.17)

Or equivalently,

$$f(k.t) = \frac{(1-k)\mu^{1-\frac{1}{k}}}{(\mu - kt)^{2-\frac{1}{k}}}.$$ (3.18)

The familiar results are found in the various limiting cases (k=0, ½, 1). For instance,

$$\text{Lim}_{k \to 0} f(k.t) = \frac{1}{\mu} e^{-\frac{t}{\mu}}$$ (3.19)

for the exponential distribution;
  and

$$f\left(\frac{1}{2} ; t \right) = \frac{1}{2\mu}$$ (3.20)

for the uniform distribution.

And for the Dirac distribution ($k = 1$), $f(k.t)$ is equal to 0 for $t < \mu$ and has a pole ($f(k.t) \to \infty$) at $t = \mu$.

There follows from (3.17) that the variance of the TTF, σ^2, is given by

$$\sigma^2 = \int_0^{\frac{\mu}{k}} \frac{(1-k)\mu^{1-\frac{1}{k}}}{(\mu - kt)^{2-\frac{1}{k}}} (t - \mu)^2 \, dt.$$ (3.21)

This integral can be calculated in closed form, which results in

$$\left(\frac{\sigma}{\mu} \right)^2 = \frac{1-k}{1+k}.$$ (3.22)

Thus the square of the coefficient of variation is expressed in a simple way as a function of the average RUL loss rate k.

Equivalently, by inverting:

$$k = \frac{1 - \left(\frac{\sigma}{\mu} \right)^2}{1 + \left(\frac{\sigma}{\mu} \right)^2}.$$ (3.23)

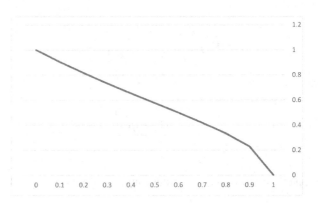

FIGURE 3.2　Coefficient of Variation as a function of RUL loss rate parameter k.

Equations (3.22) and (3.23) show that the coefficient of variation of the TTF decreases from 1 (exponential distribution) to 0 (Dirac distribution) as the RUL loss rate increases from 0 to 1, as illustrated in Figure 3.2.

For instance, for the uniform distribution ($k = \frac{1}{2}$), the coefficient of variation is found to be equal to $\frac{1}{\sqrt{3}}$, a well-known result.

3.4.2　Coefficient of Variation of the RUL

Let us first calculate the variance of the RUL, denoted by σ^2_{RUL}.
By definition,

$$\sigma^2_{RUL} = E\left(RUL^2\right) - E\left(RUL\right)^2 = E\left(RUL^2\right) - V^2\left(t\right). \tag{3.24}$$

Then, denote s the instant of failure, or end of life. The RUL at time t is given by

$$RUL\left(t\right) = s - t.$$

Therefore

$$E\left(RUL^2\right) = \int_0^\infty \left(s - t\right)^2 f_t\left(s\right)ds, \tag{3.25}$$

where $f_t\left(s\right)$ denotes the conditional probability density function of the failure time s ($>t$) under the condition that there is no failure up to time t.

$$f_t\left(s\right) = -\frac{dR_t}{ds} \tag{3.26}$$

with $R_t\left(.\right)$ the conditional reliability function at time t, i.e.

$$R_t(s) = P[R > t + s | T > t] = \frac{R(t+s)}{R(t)}. \tag{3.27}$$

Hence

$$f_t(s) = \frac{f(t+s)}{R(t)} \tag{3.28}$$

and (3.25) is equivalent to

$$E(\text{RUL}^2) = \frac{1}{R(t)} \int_0^\infty (s-t)^2 f(t+s) ds. \tag{3.29}$$

Integration by parts then shows that this is equivalent to

$$E(\text{RUL}^2) = 2 \int_0^\infty s R_t(s) ds \tag{3.30}$$

and, using (3.24), therefore,

$$\sigma^2_{\text{RUL}}(t) = 2 \int_0^\infty s R_t(s) ds - V^2(t). \tag{3.31}$$

This result is completely general (see Banjevic 2009). In the case of the family of distributions under study, we have, from (3.7):

$$R_t(s) = \left(\frac{1 - \dfrac{k(t+s)}{\mu}}{1 - \dfrac{kt}{\mu}} \right)^{\frac{1-k}{k}} = \left(1 - \frac{ks}{\mu - kt} \right)^{\frac{1-k}{k}} \quad \text{for } 0 \le s \le \frac{\mu}{k} - t \tag{3.32}$$

$$R_t(s) = 0 \quad \text{for} \quad s > \frac{\mu}{k} - t$$

There follows:

$$\int_0^\infty s R_t(s) = \frac{(\mu - kt)^2}{1+k} \tag{3.33}$$

and therefore, using (3.31),

$$\sigma^2_{\text{RUL}}(t) = 2 \frac{(\mu - kt)^2}{1+k} - (\mu - kt)^2 = (\mu - kt)^2 \frac{1-k}{1+k}.$$

Therefore the square of the coefficient of variation of the RUL, at time t, is

$$CV^2 = \left(\frac{\sigma^2_{\text{RUL}}(t)}{V(t)^2} \right) = \frac{1-k}{1+k} \tag{3.34}$$

which is independent of t.

This result proves and quantifies the statement made in the introduction, at least for the family of distributions with MRL linear in time; the RUL is time-dependent and stochastic but there is an opposition between those two characteristics, which is captured by the k coefficient: the higher k, the more time-dependent the RUL is (because the slope of the MRL is steeper) but the less variability about the mean it exhibits (because the coefficient of variation is smaller).

At one extreme—the exponential distribution, there is no time dependence of the mean (MRL is constant) but highest variability of the RUL about the mean (coefficient of variation equal to 1); at the other extreme—the Dirac distribution, there is maximum time dependence of the mean (the slope of the MRL is −1) but no variability, no fluctuation of the RUL about it (coefficient of variation equal to 0).

Clearly, prognostics and predictive maintenance are interesting for assets which undergo progressive degradations before failures; and not at all when failures are completely sudden (which is the case of the exponential distribution).

In general, therefore, one could say that $k \geq \dfrac{1}{2}$ characterizes the assets that are good candidates for prognostics.

For instance, in the crack growth example mentioned earlier, a quasi-deterministic model (such as the Paris–Erdogan law, Paris 1963) is quite realistic, so that $k \approx 1$ and that type of physical degradation is quite suitable for prognostics (uncertainties can be introduced for instance by a Wiener process, as will be shown in Chapter 5, but still with a reasonably small coefficient of variation). On the other hand, many electronics devices, characterized by sudden failures, and therefore by an exponential TTF ($k=0$), are totally unsuited for predictive maintenance. (This remarks does not apply to all electronics devices by far; for instance, capacitors or LEDs do degrade and their physics of degradations has been well studied, see, e.g., Pecht 2008; an LED example will be presented in Chapter 8).

Note: if $k<0$, (3.34) is still valid provided $k > -1$, and it shows that, in those cases of reliability growth, the coefficient of variation is greater than 1, making prognostics even more difficult.

3.5   CONFIDENCE INTERVAL FOR RUL AND RELATION TO RUL LOSS RATE

Let us now determine the confidence interval of the RUL at time t, with confidence level $1 - \alpha$.

To that end, define $s+$ and $s-$ by

$$R_t\left(s^+\right) = \frac{\alpha}{2} \qquad (3.35)$$

$$R_t\left(s^-\right) = 1 - \frac{\alpha}{2}. \qquad (3.36)$$

Then if T denotes the random variable "end of life", the conditional probability of survival, after t, for a time s greater than s^- but less than s^+, is:

$$P\left(t+s^- < T < t+s^+ \mid T > t\right) = 1 - \alpha, \qquad (3.37)$$

which is equivalent to stating that the confidence interval for T at a confidence level $(1-\alpha)$ is given by: $(t + s^-; t + s^+)$.

In other words, there is a probability $(1-\alpha)$ that $s^- < \mathrm{RUL}(t) < s^+$, i.e., the $(1-\alpha)$ level confidence interval for RUL(t) is $(s^-; s^+)$.

Of course, s^- and s^+ are generally functions of t.

From (3.32), (3.35) and (3.36), one then obtains the lower bound s^- and the upper bound s^+ of the confidence interval, which shows that s^- and s^+ are in fact linear functions of t.

$$s^+ = \left(\frac{\mu}{k}-t\right)\left[1-\left(\frac{\alpha}{2}\right)^{\frac{k}{1-k}}\right]. \qquad (3.38)$$

$$s^- = \left(\frac{\mu}{k}-t\right)\left[1-\left(1-\frac{\alpha}{2}\right)^{\frac{k}{1-k}}\right]. \qquad (3.39)$$

Confidence intervals for the RUL are illustrated in Figures 3.3 and 3.4, for two different values of the k parameter: k=0.6 in Figure 3.3 and k= 0.9 in Figure 3.4. It can be seen that, for the higher value of k (k=0.9), the confidence interval is narrower.

In the deterministic case $k = 1$, it is verified that, as expected, the confidence interval reduces to a point: there is no uncertainty around RUL(t), which is equal to (μ–t).

And, for $k = 0$, the known result for the exponential function is found:

$$s^+ = -\mu\ln\left(\frac{\alpha}{2}\right) \qquad (3.40)$$

$$s^- = -\mu\ln\left(1-\frac{\alpha}{2}\right). \qquad (3.41)$$

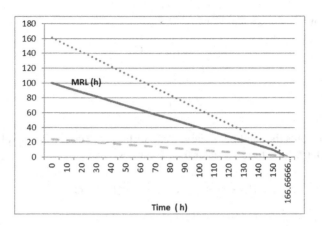

FIGURE 3.3   Confidence interval for RUL as a function of time ($k = 0.6$).

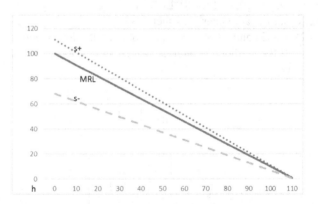

FIGURE 3.4   Confidence interval for RUL as a function of time ($k = 0.9$).

In that special case, the confidence interval does not change with time, just as the RUL.

It can be verified that the width of the confidence interval, i.e. the difference $(s^+ - s^-)$, decreases linearly with time (except in the exponential case and the deterministic case, when it is constant).

Indeed:

$$s^+ - s^- = \left(\frac{\mu}{k} - t\right)\left[\left(1 - \frac{\alpha}{2}\right)^{\frac{k}{1-k}} - \left(\frac{\alpha}{2}\right)^{\frac{k}{1-k}}\right].$$

(3.42)

And also, this width is a (nonlinear) decreasing function of k, as illustrated by Figure 3.5. It ranges from a maximum value for $k = 0$ (exponential distribution case) to a minimum value of 0 for $k = 1$ (deterministic case).

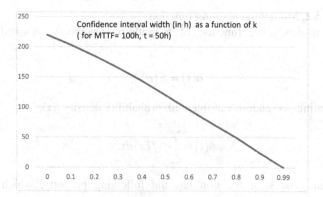

FIGURE 3.5   Width of RUL confidence interval as a function of parameter k.

This is a further illustration of the earlier statement: the faster the degradation proceeds (higher k), the less uncertainty there is on the RUL (lower CV).

This property is intuitive but it is reassuring that the mathematics confirm it.

3.6   HIGHER-ORDER MOMENTS AND MOMENT-GENERATING FUNCTION

3.6.1   MOMENT-GENERATING FUNCTION

It is sometimes useful to know the higher-order centered moments of a probability distribution, as they can all be generated by a compact moment-generating function. This notion will be used for the higher-order moments of the TTF, in Chapter 9 in the context of maintenance policy.

Here we derive a general expression for the nth order centered moment for the TTF distributions with MRL linear in time, the subject of this chapter.

Definition 3.1 (see e.g. Ross 2010). The nth order centered moment of the random variable X is defined by

$$\alpha_n = \mathrm{E}\left(X^n\right) \quad n = 0, 1, 2, \ldots. \tag{3.43}$$

In particular, the mathematical expectation is given by

$$\alpha_1 = E\left(X\right) = \mu. \tag{3.44}$$

The variance σ^2 is expressed in terms of the first-order and second-order centered moments:

$$\sigma^2 = E\left(X^2\right) - E\left(X\right)^2 = \alpha_2 - \alpha_1^2 \tag{3.45}$$

Definition 3.2. Moment-generating function

The moment-generating function of a random variable X, denoted $\phi(t)$, is defined by

$$\phi(t) = E\left(e^{tX}\right). \tag{3.46}$$

If X is a continuous random variable with probability density $f(x)$,

$$\phi(t) = \int_{-\infty}^{+\infty} e^{tx} f(x) \, dx. \tag{3.47}$$

The moment-generating function has the following property, which justifies its name:

$$E\left(X^n\right) = \phi^{(n)}(0), \tag{3.48}$$

where $\phi^{(n)}$ denotes the nth-order derivative of $\phi(t)$.

This property can be proved easily from the definition (3.46).

For instance,

$$\phi^{(1)}(t) = \frac{dE\left(e^{tX}\right)}{dt} = E\left(\frac{de^{tX}}{dt}\right) = E\left(Xe^{tX}\right) \text{ and therefore } \phi^{(1)}(0) = E(X)$$

and $\phi^{(n)}(t)$ can be obtained similarly by n successive differentiations (Ross 2010).

When the series converges, the moment-generating function can be expressed as a Taylor–McLaurin series in terms of the centered moments:

$$\phi(s) = \sum_{n=0}^{n=\infty} \frac{\alpha_n}{n!} s^n. \tag{3.49}$$

From (3.46), it is seen that $\phi(s)$ is the Laplace transform of the variable X, to within the sign: the Laplace transform is given by

$$\phi(-s) = E\left(e^{-sX}\right).$$

There follows the important property that the moment-generating function determines the distribution.

Also, the moment-generating function of a sum of independent random variables is the product of their respective moment-generating functions.

This property is useful to determine the distribution of a sum of random variables.

Sometimes the characteristic function is defined, as the Fourier transform (see e.g. Karlin 1975).

Definition 3.3. The characteristic function of the random variable X is defined as

$$\Phi(s) = E\left(e^{isX}\right) \qquad (3.50)$$

with i denoting the imaginary unit. It has the same properties as the moment-generating function.

3.6.2 MOMENT-GENERATING FUNCTION FOR SPECIAL TTF FAMILY

Let T denote the TTF for a distribution with an MRL linear function of time, and $f(t)$ the probability density of T.

Let us calculate the nth order centered moment α_n.

$$\alpha_n = \int_0^\infty t^n f(t)\, dt = \int_0^\infty t^n \frac{(1-k)}{\mu}\left(1 - \frac{kt}{\mu}\right)^{\frac{1}{k}-2} dt. \qquad (3.51)$$

By induction, it is shown easily from (3.51) that

$$\alpha_n = \frac{n\mu}{k(n-1)+1}\alpha_{n-1} \quad n = 1,2,3,\dots \qquad (3.52)$$

with

$$\alpha_0 = 1. \qquad (3.53)$$

There follows:

$$\alpha_n = \frac{n!\,\mu^n}{\prod_{j=1}^{j=n-1}(jk+1)}. \qquad (3.54)$$

Using the property of the Euler gamma function,

$$\Gamma(n+1) = n!$$

for positive integer n, (3.54) can also be written as

$$\alpha_n = \frac{n!\,\mu^n}{k^n}\frac{\Gamma\left(\dfrac{1}{k}\right)}{\Gamma\left(\dfrac{1}{k}+n\right)}. \qquad (3.55)$$

For instance, for $n = 1$, (3.54) gives, using the property that $\Gamma(x + 1) = x\Gamma(x)$,

$$\alpha_1 = \mu \qquad (3.56)$$

and for $n = 2$,

$$\alpha_2 = \frac{2\mu^2}{k^2} \frac{\Gamma\left(\frac{1}{k}\right)}{\Gamma\left(\frac{1}{k}+2\right)} = \frac{2\mu^2}{k^2} \frac{\Gamma\left(\frac{1}{k}\right)}{\left(\frac{1}{k}+1\right)\Gamma\left(\frac{1}{k}+1\right)} = \frac{2\mu^2}{k^2} \frac{\Gamma\left(\frac{1}{k}\right)}{\left(\frac{1}{k}+1\right)\frac{1}{k}\Gamma\left(\frac{1}{k}\right)} = \frac{2\mu^2}{1+k}.$$

(3.57)

From (3.57) and (3.56), the relation between the k parameter and the coefficient of variation is found again:

$$\sigma^2 = \alpha_2 - \left(\alpha_1\right)^2 = \frac{2\mu^2}{1+k} - \mu^2 = \frac{1-k}{1+k}\mu^2.$$

(3.58)

A general expression for the moment-generating function is then obtained from (3.49):

$$\phi(s) = \sum_{n=0}^{n=\infty} \frac{\alpha_n}{n!} s^n = \dots$$

(3.59)

and the Laplace transform of the TTF distribution is

$$L(s) = \phi(-s).$$

Let us take a look at the special limiting cases.

For $k = 0$

$$\phi(s) = \sum_{n=0}^{n=\infty} (\mu s)^n = \frac{1}{1-\mu s} \quad \left(\text{for } s < \frac{1}{\mu}\right).$$

To see this, it is best to use the formulation given in (3.54).

Therefore

$$L(s) = \frac{1}{1+\mu s}$$

(3.60)

which is indeed the Laplace transform of the exponential distribution with mean μ.

For $k = 1$,

$$\phi(s) = \sum_{n=0}^{n=\infty} \frac{(\mu s)^n}{n!} = e^{\mu s}$$

which results directly from (3.55). Therefore

$$L(s) = e^{-\mu s}$$

(3.61)

and this is indeed the Laplace transform of the "Dirac" function, i.e., an impulse at $t = \mu$.

For $k = \frac{1}{2}$,

$$\phi(s) = \sum_{n=0}^{n=\infty} \frac{(\mu s)^n}{\prod_{j=1}^{j=n-1}\left(\frac{j}{2}+1\right)n!} = \frac{e^{2\mu s}-1}{2\mu s} \qquad (3.62)$$

and therefore

$$L(s) = \frac{1-e^{-2\mu s}}{2\mu s} \qquad (3.63)$$

which is indeed the Laplace transform of the function $f(t)$ defined by

$$f(t) = \begin{cases} \dfrac{1}{2\mu}, & 0 \le t \le 2\mu \\ 0, & t < 0 \text{ or } t > 2\mu \end{cases} \qquad (3.64)$$

in other words, the probability density function of the uniform distribution over $(0, 2\mu)$, with mean μ.

Those are just verifications on limiting cases of the general expression (3.59) for the moment-generating function of a TTF distribution belonging to special family studied in this chapter, i.e., the distributions with MRL linear in time.

3.7 CUMULATIVE HAZARD FUNCTION

The cumulative hazard function is obtained from the failure rate as follows (cf. Chapter 2):

$$\Lambda(t) = \int_0^t \lambda(s)ds = \int_0^t \frac{(1-k)}{(\mu-ks)}ds = (1-k)\int_0^t \frac{1}{(\mu-ks)}ds = \frac{1-k}{k}\left[-\log\left(1-\frac{k}{\mu}t\right)\right].$$

$$(3.65)$$

In limiting cases, the expected results are verified, using the Taylor series expansion of the function $\log(1+x)$ for $|x| < 1$.

$$\lim_{k\to 0} \Lambda(t) = \frac{t}{\mu}$$

$$\lim_{k\to 1} \Lambda(t) = 0 \ t < \mu$$

and, for $k = 1/2$, $\Lambda(t) = -\log\left(1 - \dfrac{t}{2\mu}\right)$.

These results will be useful in Chapter 9 on maintenance.

3.8   EXERCISES

1. For train rolling stock equipment, mean distance to failure is the usual reliability measure. Assume a degradation phenomenon in a train item of equipment progresses proportionally to mileage run, i.e. the MRL is a linear function of mileage; and, after each mile (or km) of operation, the RUL is reduced by one quarter of a mile (or km) on average.

   Assume further that the mean distance to failure (MDTF) is equal to 1 million km.
   a. What is the MRL after 500 000 km?
   b. What is the probability that the item fails before 300 000 km?
   c. Calculate the failure rate at 500 000 km.

2. For the example of Exercise 1, what is the coefficient of variation of the RUL?
   Determine the bounds of the 80% confidence interval for the RUL, at 500 000 km.

3. Consider a different item of equipment in the rolling stock, which also degrades proportionally to mileage run, but faster: its RUL is reduced on average by three quarters of a mile after each mile run. Assuming it has the same MDBF as the first item, perform Exercise 1 and Exercise 2 again with those new assumptions. Compare the two items. Which one would be the more appropriate candidate for predictive maintenance?

4. In statistics, the skewness (Pearson's moment definition of skewness) of a random variable X is defined by the following index:

$$sk(X) = E\left[\frac{(X-\mu)^3}{\sigma^3}\right]$$ in terms of mean μ and standard deviation σ.

   (It measures how skewed the distribution of X is as compared to the normal distribution.)
   Calculate the skewness of the TTF distribution in the above two examples
   (Hint: use the centered moment of degree 3, applying the formula of Section 3.6.2, as well as the standard deviation.)
   Which one of the two is more skewed?
   In general, show that the skewness depends only on k (not μ) and study the impact of k. What is the skewness of the exponential distribution?

BIBLIOGRAPHY

Bajenesco, T., "Degradation and Reliability Problems of Optocouplers", *Proceedings of the 16th Annual Semiconductor Conference CAS '93, October 12–17, 1993*, Sinaia, Romania, pp. 275–280.

Black, J.R., "Mass Transport of Aluminum by Momentum Exchange with Conducting Electrons," 6th Annual Reliability Physics Symposium (IEEE), 1967, pp. 148–159, doi:10.1109/IRPS.1967.362408

Banjevic, D., "Remaining Useful Life in Theory and Practice", *Metrika*, 69, 337–349, 2009.

Dersin, P., "The Class of Life Time Distributions with a Mean Residual Life Linear in Time—Application to PHM" in *Safety & Reliability: Safe Societies in a Changing World*, Eds S. Haugen, A. Barros et al., London, CRC Press, 1093–1099, 2018.

Karlin, S., & Taylor, H.M., *A First Course in Stochastic Processes*. London, Academic Press, 1975.

Paris, P. C., Erdogan, F., "A Critical Analysis of Crack Propagation Laws", *Journal of Basic Engineering*, 85(4): 528–533, 1963. doi:10.1115/1.3656900

Pecht, M., *Prognostics & Health Management of Electronics*. Hoboken, NJ, John Wiley & Sons, 2008, ISBN: 978-0-470-38584-5

Ross, S.M., *Introduction to Probability Models*, 10th Ed., Burlington, MA, Academic Press, Elsevier, 2010.

4 Generalization to an MRL Piecewise-Linear Function of Time

4.1 RELIABILITY FUNCTION, MRL, AND FAILURE RATE

In this short chapter, we now address the case when MRL is a piecewise-linear function of time.

Consider first the case of a time-to-failure characterized by an MRL that consists of two linear segments, i.e. a piecewise-linear function with two pieces, as shown in Figure 4.1. It is then possible to generalize to any number of segments.

In that example, the MRL has a change of slope at time $T_1 = 40h$ and is defined as follows:

$$V(t) = \mu - k_1 t \quad \text{for } 0 \le t \le T_1 \tag{4.1}$$

$$\mu - k_1 T_1 - k_2 (t - T_1) \text{ for } T_1 \le t \le T_2 \text{ with } V(T_2) = 0 \tag{4.2}$$

From (4.1) and (4.2),

$$\mu - k_1 T_1 - k_2 (T_2 - T_1) = 0. \tag{4.3}$$

The example of Figure 4.1 can be seen as a system which undergoes a degradation process with a step change at time T_1. For instance, a new failure mechanism could kick in at time $T_1 = 40h$. As a result, the absolute value of the slope of the average RUL loss changes abruptly from $k_1 = 0.4$ before time T_1 to $k_2 = 0.9$ after that time.

It is assumed that $\mu = 100h$, which, consistently with (4.3), implies that $T_2 = 133.33h$.

From (4.1), the reliability function is derived, using (2.18) of Chapter 2:

$$\text{For } 0 < t < T_1, R(t) = \left(1 - \frac{k_1}{\mu} t\right)^{\frac{1}{k_1} - 1}. \tag{4.4a}$$

DOI: 10.1201/9781003250685-4

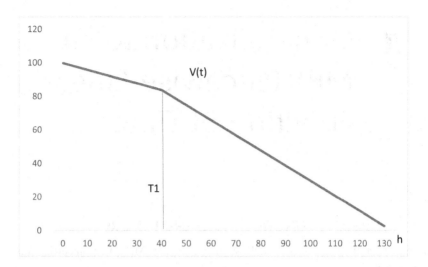

FIGURE 4.1   Piecewise-linear MRL (two segments in this example).

For $T_1 < t < T_2$

$$R(t) = \frac{1}{1 - \frac{k_1}{\mu}T_1 - \frac{k_2}{\mu}(t - T_1)} \left(1 - \frac{k_1}{\mu}T_1\right)^{\frac{1}{k1}} \left[1 - \frac{(t - T_1)}{\left(\frac{\mu}{k_2} - \frac{k_1}{k_2}T_1\right)}\right]^{\frac{1}{k2}} . \quad (4.4b)$$

The reliability function $R(t)$ is plotted against time in Figure 4.2.

From those expressions, it is possible to derive a confidence interval for the RUL, in the same manner as was done in Chapter 3 for the linear case.

Similarly the probability density function $f(t)$ can be derived from (4.4), and therefore also the failure rate $\lambda(t)$.

The failure rate can be obtained in two ways: either as just stated, or directly from $V(t)$ as per (2.11) of Chapter 2. The latter method is faster. It leads to

$$\lambda(t) = \frac{1 + \frac{dV}{dt}}{V(t)} \quad (4.5)$$

which, with $V(t)$ given by (4.1), yields:

$$\lambda(t) = \frac{1 - k_1}{\mu - k_1 t} \quad \text{for } 0 \le t \le T_1 \quad (4.6a)$$

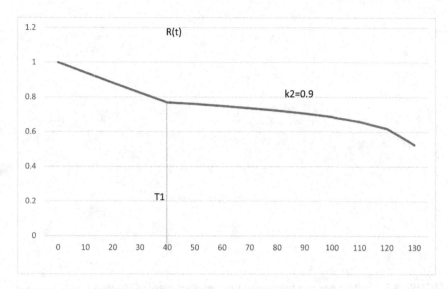

FIGURE 4.2 Reliability function in the example of Figure 4.1.

$$\lambda(t) = \frac{1-k_2}{\mu - k_1 T_1 - k_2 (t - T_1)} \quad \text{for} \quad T_1 < t < T_2. \tag{4.6b}$$

It can be verified that the same expression is arrived at by differentiating $R(t)$.

The failure rate is plotted against time in Figure 4.3.

Note that the failure rate has a discontinuity at T_1. The MRL is not the same as the MRL which would correspond to a new failure rate that adds to the initial failure rate from time T_1 onwards.

Indeed, that situation would be described by the following failure rate function:

$$\lambda(t) = \frac{1-k_1}{\mu_1 - k_1 t} \quad \text{for} \quad 0 \le t \le T_1 \tag{4.7a}$$

and

$$\lambda(t) = \frac{1-k_1}{\mu_1 - k_1 t} + \frac{1-k_2}{\mu_2 - k_2 t} \quad \text{for} \quad t > T_1 \tag{4.7b}$$

But, in that case, according to (2.11) or (4.5), $V(t)$ would not be a piecewise-linear function, as it would verify (4.5) with $\lambda(t)$ given by (4.7).

The above result is easily generalized to any number of segments (instead of two in the example).

Thus the generality is considerably extended, since any continuous function can be approximated by a piecewise-linear function.

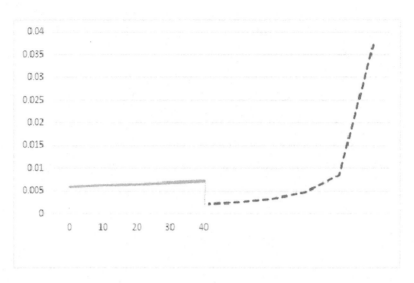

FIGURE 4.3   Failure rate as a function of time in the example of Figure 4.1.

4.2 EXERCISES

1. In the two-segment piecewise-linear function described in this chapter, build a confidence interval for the RUL.

2. Express the reliability function for a time-to-failure characterized by a MRL with *three* linear segments, i.e.:

$$V(t) = \mu - k_1 t \quad 0 \leq t \leq T_1$$
$$\mu - k_1 T_1 - k_2 \left(t - T_1\right) \quad T_1 \leq t \leq T_2$$
$$\mu - k_1 T_1 - k_2 \left(T_2 - T_1\right) - k_3 \left(t - T_2\right) \quad T_2 \leq t \leq T_3$$

3. For the case of Exercise 2, derive an expression for the failure rate as a function of time.

4. Consider a degradation process characterized by (i) a first period of 720h (approximately one month), during which the average loss rate is 50%, in other words one day over two days of operation; (ii) a second period of 720 h, during which the average RUL loss rate is one day per day. Calculate:

   a. The probability of reaching 720h without failure;
   b. The probability of reaching 1080h without failure;
   c. The mean residual life after 720h and after 1080h.

5. Express the reliability function $R(t)$ for an asset that first exhibits a constant failure rate during a period $(0, T_1)$, and subsequently a deterministic failure pattern (Dirac) during a period (T_1, T_2).

BIBLIOGRAPHY

Ross, S.M., *Introduction to Probability Models*. Burlington, MA, Elsevier, 2010.

5 Generalization to a Wide Class of Lifetime Distributions

5.1 INTRODUCTION: GENERALIZATION METHOD

In Chapter 3, we studied the properties of an important family of lifetime distributions, characterized by the fact that their MRL is a linear function of time.

Then, in Chapter 4, it was shown how those properties can be generalized to the case of an MRL which is a piecewise-linear function of time.

However, one would like to relate properties of well-known lifetime distributions, such as Weibull or Gamma, to the above-mentioned results. This is the objective of this chapter.

The method that will be utilized for that purpose is that of a time transformation: it will be shown that it is possible to define a (generally nonlinear) transformation of the time variable in such a way that, in the transformed time variable, the MRL is a linear function of time. Then, in that "transformed space", the properties presented in Chapter 3 hold and, by applying the time transformation backward, they can be converted into properties in the initial space. In particular, confidence intervals for the RUL can be constructed by that method. The properties of the average RUL loss rate can also be characterized for each distribution type.

This analysis can be performed for a wide variety of lifetime distributions, which are in fact two-parameter distributions, where one parameter is a scale parameter, another is a shape parameter (with sometimes a shift parameter as well).

5.2 NONLINEAR TIME TRANSFORMATION

It is desired to find an invertible transformation $g(.)$ of the nonnegative half line $[0,\infty)$:

$$t \rightarrow t' = g(t) \tag{5.1}$$

such that, in the "transformed time" t', the time to failure has an MRL which is a linear function (of the transformed time t').

DOI: 10.1201/9781003250685-5

We shall impose *invariance of the reliability function* under that transformation:

$$R'(t') = R(t), \tag{5.2}$$

Let T' denote the random variable $g(T)$:

$$T' \equiv g(T). \tag{5.3}$$

The solution is then given by the following theorem.

Theorem 5.1

Let T denote the random variable "time to failure" (or "lifetime"), with reliability function $R(t)$, expectation μ and variance σ^2.

$$E(T) = \mu \tag{5.4}$$

$$\mathrm{Var}(T) = \sigma^2. \tag{5.5}$$

Then the transformation $g()$ defined below (5.6) transforms T into a random variable T' with an MRL that is a linear function of t', and with the same mean and variance as T.

Definition of g(t):

$$g(t) = \frac{\mu}{k}\left[1 - R(t)^{\frac{k}{1-k}}\right] \tag{5.6}$$

where

$$\mu \equiv E(T) \tag{5.7}$$

and

$$k \equiv \frac{1 - \left(\dfrac{\sigma}{\mu}\right)^2}{1 + \left(\dfrac{\sigma}{\mu}\right)^2} \tag{5.8}$$

or equivalently

$$\sigma^2 = \mu^2 \frac{1-k}{1+k}. \tag{5.9}$$

Proof

The definition of $g(t)$ implies that

$$R(t) = \left[1 - k\frac{g(t)}{\mu}\right]^{\frac{1-k}{k}}$$

(5.10)

$$0 \le t < \infty.$$

But, given (5.1) and (5.2), one can therefore write:

$$R'(t') = \left(1 - k\frac{t'}{\mu}\right)^{\frac{1-k}{k}}$$

(5.11)

$$0 \le t' < \frac{\mu}{k},$$

which shows that the probability distribution of T' is that of a lifetime with an MRL linear in time t', with slope parameter k:

$$V'(t') = \mu - kt'.$$

(5.12)

Now let us show that $E(T') = \mu$, and Var $(T') = \sigma^2$, i.e., the mean and variance of the transformed time to failure T' are equal respectively to the mean and variance of T, the time to failure in the initial referential.

In fact we prove a more general result on the centered moments of $T' = g(T)$, the transformed time.

Let α'_n denote the nth order centered moment of T', with distribution given by (5.11); denoting the density by $f(t) = -\dfrac{dR}{dt}$ as usual,

$$\alpha'_n = E\left[(T')^n\right] = \int_0^{\frac{\mu}{k}} s^n \mathrm{f}'(s)\,ds = \int_0^\infty g(t)^n\, \mathrm{f}'(g(t)).\frac{dg}{dt}(t)\,dt.$$

Integrating by parts,

$$\int_0^{\frac{\mu}{k}} s^n f'(s)\,ds = n\int_0^{\frac{\mu}{k}} s^{n-1} R'(s)\,ds = n\int_0^\infty g(t)^{n-1} R'(g(t)).\frac{dg}{dt}\,dt.$$

Then, using the fact that

$$R'(g(t)) = R'(t') = R(t),$$

$$\alpha'_n = n \int_0^\infty g(t)^{n-1} R(t) \frac{dg}{dt} dt. \tag{5.13}$$

Now, from the definition (5.6) of $g(t)$,

$$\frac{dg}{dt} = \frac{\mu}{1-k} R(t)^{\frac{2k-1}{1-k}} f(t).$$

Therefore

$$\alpha'_n = n \frac{\mu}{1-k} \left(\frac{\mu}{k}\right)^{n-1} \int_0^\infty \left[1 - R(t)^{\frac{k}{1-k}}\right]^{n-1} R(t)^{\frac{k}{1-k}} f(t) dt$$

$$= \frac{n\mu^n}{k^{n-1}(1-k)} \int_0^1 \left[1 - R(t)^{\frac{k}{1-k}}\right]^{n-1} R(t)^{\frac{k}{1-k}} dR(t)$$

or,

$$\alpha'_n = \frac{n\mu^n}{k^{n-1}(1-k)} \int_0^1 \left[1 - x^{\frac{k}{1-k}}\right]^{n-1} x^{\frac{k}{1-k}} dx. \tag{5.14}$$

With the change of variable $u = x^{\frac{k}{1-k}}$ (provided $0 < k < 1$), this reduces to

$$\alpha'_n = n \frac{\mu^n}{k^n} \int_0^1 u^{\frac{1}{k}-1} (1-u)^{n-1} du$$

which can be expressed in terms of the Euler gamma function Γ:

$$\alpha'_n = n \frac{\mu^n}{k^n} \frac{\Gamma\left(\frac{1}{k}\right)\Gamma(n)}{\Gamma\left(\frac{1}{k}+n\right)}.$$

As n is an integer,

$$\Gamma(n) = (n-1)!$$

and there follows that

$$\alpha'_n = n! \left(\frac{\mu}{k}\right)^n \frac{\Gamma\left(\frac{1}{k}\right)}{\Gamma\left(\frac{1}{k}+n\right)}. \tag{5.15}$$

It is—not surprisingly—the same expression that was encountered in Chapter 3 (Section 3.6.2) for the nth order centered moment of a time-to-failure distribution with MRL linear in time, with parameters μ and k.

In particular, (5.13) yields:

- for $n = 1$, $\alpha'_1 = \mu$, i.e., $E(T') = E(T)$
- for $n = 2$, $\alpha'_2 = \dfrac{2\mu^2}{1+k}$

and therefore, the variance of T' is equal to

$$\text{Var}(T') = E(T'^2) - E(T')^2 = \frac{2\mu^2}{1+k} - \mu^2 = \frac{1-k}{1+k}\mu^2 = \sigma^2 = \text{Var}(T)$$

qed.

Discussion

1. It is also verified from (5.10) and (5.11) that the reliability function is invariant under the transformation, i.e. (5.2) holds. One can also see immediately that, if the initial distribution belongs to the special family of distributions with an MRL linear in time, the transformation is the identity function:

$$g(t) = t.$$

While time ranges from 0 to infinity in the initial distribution, the "transformed time" t' ranges from 0 to μ/k in the image space since, from (5.6),

$$g(0) = 0$$

$$\lim_{t \to \infty} g(t) = \frac{\mu}{k}.$$

2. Important remark: the centered moments of the initial distribution T are generally not the same as for the transformed distribution $T' = g(T)$; except for $n = 1$ and $n = 2$.

In other words, in general:

$$\alpha_n \equiv \int_0^\infty t^n f(t)dt \neq \int_0^{\frac{\mu}{k}} s^n f'(s)ds \equiv \alpha'_n \quad \text{for} \quad n > 2.$$

For $n = 1$,

$$\alpha_1 = \int_0^\infty t f(t)dt = \int_0^\infty R(t)dt \quad \text{(integrating by parts)}.$$

Then, since $R(t) = R'(t')$, the right-hand side is equal to $\int_0^{\frac{\mu}{k}} R'(t')dt' = \mu = \alpha_1'$.

For $n = 2$,

$$\alpha_2 = \int_0^\infty t^2 f(t)dt = 2\int_0^\infty sR(s)ds.$$

Integration by parts and using $R'(t') = R(t)$ leads, as shown in the proof of Theorem (5.1), to

$$\alpha_2' = \int_0^{\frac{\mu}{k}} s^2 f' s)ds = \frac{2\mu^2}{1+k}$$

with k given by (5.8).
Therefore

$$\alpha_2' = 2\mu^2(1+k).$$

But, from (5.8),

$$1+k = 1 + \frac{1 - \dfrac{\sigma^2}{\mu^2}}{1 + \dfrac{\sigma^2}{\mu^2}}.$$

Therefore

$$1+k = \frac{2}{1 + \dfrac{\sigma^2}{\mu^2}}$$

and

$$\alpha_2' = 2\mu^2(1+k)\mu^2 + \sigma^2 = \alpha_2.$$

To see that, in general, $\alpha_n' \neq \alpha_n$ for $n > 2$, one may consider for instance the case $n = 3$, which, for $k = 1/2$ for instance, leads to $\alpha_3' = 2\mu^3$ by application of (5.15), and show by taking an example of TTF distribution, that this is in general different from α_3.

This will done later in this chapter when we examine particular distributions.

In general then, the transformation $g()$ has two parameters, μ and k, which can be determined by (5.7) and (5.8) from the first- and second-order moments of the distribution.

There are not enough degrees of freedom to select additional parameters, such as higher-order moments; they are determined by the choice of μ and k.

Thus the method presented in Theorem 5.1 works for two-parameter distributions (or three-parameter distributions when the third parameter is a shift).

3. In the proof of Theorem 5.1, it was assumed that $0 < k < 1$. Let us now look at the limiting cases $k = 0$ and $k = 1$.

In the case $k = 0$ (exponential distribution), α'_n is calculated directly from (5.13), with $R(t) = e^{-\lambda t}$, $(\lambda = \dfrac{1}{\mu})$ and $g(t) = t$; which gives

$$\alpha'_n = \frac{n!}{\lambda^n} = n!\mu^n.$$

In the case $k = 1$ ("Dirac"), $R(t)$ is a step function with step at $t = \mu$:

$$R(t) = \begin{cases} 1 & t < \mu \\ 0 & t \geq \mu \end{cases}$$

and again $g(t) = t$, and (5.13) gives $\alpha'_n = \mu^n$.

The results for those two limiting cases are also found from (5.15) by letting k go to 0 and 1, respectively.

Those special cases had been encountered in Chapter 3 (Section 3.6.2) when discussing the moment-generating function.

5.3 CONFIDENCE INTERVAL FOR RUL

A confidence interval for the RUL will now be constructed, by using the results of Chapter 3 and the nonlinear time transformation just introduced in Section 5.2.

Consider a time t. It is desired to build a confidence interval around RUL(t).

But, in order to draw on the known results for confidence intervals around the RUL for special distributions with an MRL linear function of time (Chapter 3), we rather reason in the 'T'- space' with T' defined by (5.3), (5.6).

If t' is some time value, then RUL(t'), the remaining useful life at time t', is the difference between the failure time, or end-of-life, t'_f, and the current time t':

$$\text{RUL}(t') = t'_f - t'. \tag{5.16}$$

Since T' has a distribution whose MRL is a linear function of time t', it is known how to find a confidence interval for RUL(t'): by using (3.38) and (3.39) of Chapter 3.

At confidence level $(1-\alpha)$ therefore, the following holds:

$$P\left[s'^{-} \leq t'_f - t' \leq s'^{+}\right] = 1 - \alpha \qquad (5.17)$$

with s'^{-} and s'^{+} given by (3.38) and (3.39), respectively, in terms of $t' = g(t)$.
Therefore one can write:

$$P\left[t' + s'^{-} \leq t'_f \leq t' + s'^{+}\right] = 1 - \alpha \qquad (5.18)$$

or,

$$P\left[g(t) + s'^{-} \leq g(t_f) \leq g(t) + s'^{+}\right] = 1 - \alpha \qquad (5.19)$$

or, in terms of the inverse function g^{-1},

$$P\left[g^{-1}\left(g(t) + s'^{-}\right) \leq t_f \leq g^{-1}\left(g(t) + s'^{+}\right)\right] = 1 - \alpha \qquad (5.20)$$

which provides a confidence interval for the end of life t_f, from which is derived the following confidence interval for the RUL:

$$g^{-1}\left(g(t) + s'^{-}\right) - t \leq \text{RUL}(t) = t_f - t \leq g^{-1}\left(g(t) + s'^{+}\right) - t \qquad (5.21)$$

with probability $(1-\alpha)$.

The bounds, s'^{-} and s'^{+}, are obtained from (3.38) and (3.39) by substituting $t' = g(t)$ for t:

$$s'^{+} = \left(\frac{\mu}{k} - g(t)\right)\left[1 - \left(\frac{\alpha}{2}\right)^{\frac{k}{1-k}}\right] \qquad (5.22)$$

$$s'^{-} = \left(\frac{\mu}{k} - g(t)\right)\left[1 - \left(1 - \frac{\alpha}{2}\right)^{\frac{k}{1-k}}\right] \qquad (5.23)$$

With a little bit of algebra, taking (5.22) and (5.23) into account, (5.21) can be expressed as stated in Theorem 5.2.

Theorem 5.2
For a lifetime distribution characterized by the transformation $g(.)$ defined by (5.6), (5.7), (5.8), the $(1-\alpha)$-level confidence interval for RUL at time t is described as follows:

$$L_\alpha(t) \leq \text{RUL}(t) \leq U_\alpha(t) \qquad (5.24)$$

with

$$L_\alpha(t) = g^{-1}\left[\frac{\mu}{k}\left[1-\left(1-\frac{\alpha}{2}\right)^{\frac{k}{1-k}}\right]+g(t)\left(1-\frac{\alpha}{2}\right)^{\frac{k}{1-k}}\right]-t \qquad (5.25)$$

$$U_\alpha(t) = g^{-1}\left[\frac{\mu}{k}\left(1-\left(\frac{\alpha}{2}\right)^{\frac{k}{1-k}}\right)+g(t)\left(\frac{\alpha}{2}\right)^{\frac{k}{1-k}}\right]-t. \qquad (5.26)$$

Example: consider an 80% confidence level: $1-\alpha = 0.80$.
Then the lower bound of the confidence interval is

$$L_{0.2}(t) = g^{-1}\left[\frac{\mu}{k}\left(1-(0.9)^{\frac{k}{1-k}}\right)+g(t)(0.9)^{\frac{k}{1-k}}\right]-t \qquad (5.27)$$

And the upper bound is given by

$$U_{0.2}(t) = g^{-1}\left[\frac{\mu}{k}\left(1-(0.1)^{\frac{k}{1-k}}+g(t)(0.1)^{\frac{k}{1-k}}\right]-t \qquad (5.28)$$

Discussion
It is seen from (5.25) and (5.26) that, as expected, the width of the confidence interval converges to 0 as $t\to\infty$. Indeed remember that

$$\text{Limit}_{t\to\infty}\, g(t) = \frac{\mu}{k}.$$

Theorem 2 is now applied to several well-known probability distributions.

5.4 APPLICATION TO WEIBULL DISTRIBUTION

5.4.1 DERIVATION OF THE NONLINEAR TRANSFORMATION

The method just outlined in Section 5.3 will now be applied to a number of frequently encountered lifetime distributions, starting with the Weibull distribution.

The Weibull distribution is extensively used in mechanics, for instance, as it was introduced by the Swedish engineer W. Weibull (1939) in connection with material strength calculations. Also, it can be shown that, if there are several failure modes that are present simultaneously, the asset will fail when the first failure mode is activated. In that way, the Weibull distribution is connected to the family of extreme value distributions (Gumbel 1958, Nachlas 2005).

The reliability function of the two-parameter Weibull distribution is characterized by a scale parameter η and a shape parameter β. It is expressed as:

$$R(t) = e^{-\left(\frac{t}{\eta}\right)^{\beta}}.$$

(5.29)

The nonlinear transformation $g(.)$ for the Weibull distribution is then obtained by applying (5.6):

$$g(t) = \frac{\mu}{k}\left[1 - e^{-\frac{k\left(\frac{t}{\eta}\right)^{\beta}}{1-k}}\right].$$

(5.30)

This equation contains two unknown parameters, μ and k, which can be determined by expressing the fact that both mean and variance remain invariant under the transformation $g(.)$.

Mean μ:

$$\mu = \eta\Gamma\left(1 + \frac{1}{\beta}\right),$$

(5.31)

where the known expression for the mean of the Weibull distribution (Birolini 2017) has been used (Γ denotes Euler's gamma function).

Variance σ^2:

$$\sigma^2 = \eta^2\left(\Gamma\left(1 + \frac{2}{\beta}\right) - \Gamma^2\left(1 + \frac{1}{\beta}\right)\right).$$

(5.32)

The relation (5.8) between k and σ^2 is used, along with (5.31) and (5.32), in order to determine k:

$$k = \frac{1 - \left(\frac{\sigma}{\mu}\right)^2}{1 + \left(\frac{\sigma}{\mu}\right)^2} = \frac{2\Gamma^2\left(1 + \frac{1}{\beta}\right)}{\Gamma\left(1 + \frac{2}{\beta}\right)} - 1.$$

(5.33)

Thus the transformation $g(.)$ is fully characterized by (5.30), (5.31), and (5.33).

Discussion

Note that the k coefficient depends only on the shape parameter β, not on the scale parameter η. It is seen from (5.33) that the special case of the exponential distribution, $\beta = 1$, corresponds to $k = 0$, as expected. At the other extreme, when the shape parameter β goes to infinity, the k coefficient converges to 1, which corresponds to the deterministic (Dirac) distribution, i.e. zero-variance.

The g function is plotted in Figure 5.1 for $\beta = 5$ (and $\eta = 100h$).

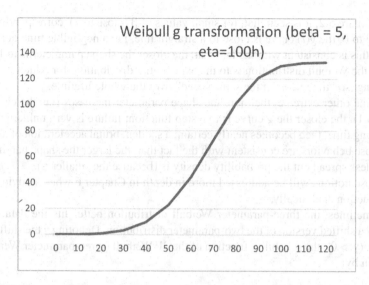

FIGURE 5.1 Nonlinear time transformation for the Weibull distribution.

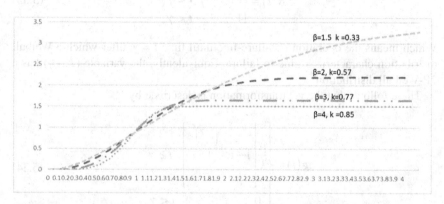

FIGURE 5.2 Sensitivity to the value of the shape factor β.

For t small, $g(t)$ remains close to 0, which means that, in the transformed space, "time is more or less standing still": the probability of a failure is very small; then, approximately in the 40h to 100h range, $g(t)$ rises quickly (faster than actual time), which corresponds to fast aging, a high average loss in RUL per unit of time; then, after the 100h milestone, the curve is nearly flat again: aging is very slow. This remark is consistent with the shape of the probability density function, which is quite concentrated, not very spread out, since β is rather large, or equivalently σ is small.

Sensitivity to the parameter k is illustrated in Figure 5.2 (for β ranging from 1.5 to 4). The plot of $g(.)$ is an S-shaped curve, ranging from $g(0) = 0$ to $g(\infty) = \mu/k$.

In Figure 5.2, it is seen that, for small values of β (close to 1), corresponding to k close to 0, the $g(t)$ curve is close to a straight line, i.e. a negligible time deformation: this is consistent with the fact that, the closer the shape parameter is to 1, the closer the Weibull distribution is to the exponential distribution, for which there is no aging, so "time flows at the same speed" over the whole lifetime.

At the other extreme, the more the shape parameter increases (i.e. the closer k gets to 1), the closer the g curve is to a step function; failure is very unlikely for a very long time, then becomes nearly certain: a sudden, brutal acceleration of aging.

Those behaviors are consistent with the fact that, the larger the shape parameter β, the less spread out the probability density is (because the smaller σ).

Those notions will be addressed more in depth in Chapter 6 when studying the $g(.)$ function analytically.

Sometimes, the three-parameter Weibull distribution better fits the data: it is simply a shifted version of the two-parameter distribution. Denoting γ the shift parameter (γ > 0), the reliability function of the 3P Weibull (three-parameter Weibull) is given by:

$$R(t) = \begin{cases} e^{-\left(\frac{t-\gamma}{\eta}\right)^{\beta}}, & t \geq \gamma. \\ 1 & , \quad t < \gamma \end{cases} \tag{5.33}$$

Which means the operation is failure-free until time $t = \gamma$, after which a Weibull distribution characterizes time to failure. Equivalently, the variable $(T - \gamma)$ has a 2P-Weibull distribution.

There follows that the $g()$ transformation is described by

$$g(t) = \begin{cases} \dfrac{\mu}{k}\left[1 - e^{-k\frac{\left(\frac{t-\gamma}{\eta}\right)^{\beta}}{1-k}}\right], & t \geq \gamma \\ 0, & t < \gamma \end{cases} \tag{5.34}$$

The standard deviation σ^2 is the same as for the 2P Weibull but the expectation differs by the shift parameter γ:

$$\mu = \gamma + \eta\Gamma\left(1 + \frac{1}{\beta}\right). \tag{5.35}$$

Therefore the k parameter is derived as before from the coefficient of variation, which results in:

$$k = \frac{1-\left(\dfrac{\sigma}{\mu}\right)^2}{1+\left(\dfrac{\sigma}{\mu}\right)^2} = \frac{1-\dfrac{\eta^2\left(\Gamma\left(1+\dfrac{2}{\beta}\right)-\Gamma^2\left(1+\dfrac{1}{\beta}\right)\right)}{\left(\gamma+\eta\Gamma\left(1+\dfrac{1}{\beta}\right)\right)^2}}{1+\dfrac{\eta^2\left(\Gamma\left(1+\dfrac{2}{\beta}\right)-\Gamma^2\left(1+\dfrac{1}{\beta}\right)\right)}{\left(\gamma+\eta\Gamma\left(1+\dfrac{1}{\beta}\right)\right)^2}}. \tag{5.36}$$

A special case, corresponding to $\beta = 1$, is the two-parameter exponential distribution:

$$R(t) = \begin{cases} e^{-\left(\frac{t-\gamma}{\eta}\right)}, & t \geq \gamma \\ 1, & t < \gamma \end{cases} \tag{5.37}$$

for which

$$g(t) = \begin{cases} \left(1+\dfrac{\gamma}{\eta}\right)(t-\gamma) \text{ if } t \gg \gamma \\ 0 \text{ if } t < \gamma \end{cases} \tag{5.38}$$

$$\mu = \eta + \gamma \tag{5.39}$$

$$k = \frac{(\eta+\gamma)^2 - \eta^2}{(\eta+\gamma)^2 + \eta^2}. \tag{5.40}$$

Those shifted distributions account for the fact that no failure will ever occur before time $t = \gamma$.

Notice that the expectation is just shifted by an amount equal to γ.

The effect on the variance is more subtle. In particular, the k coefficient is no longer equal to 0 for the shifted exponential distribution. It increases with the magnitude of the shift γ and can actually approach 1 as closely as desired is γ is large enough.

This means that, if failures are known not to occur until after a very long period (large γ), the behavior of the failure time is tantamount to deterministic.

5.4.2 Derivation of Confidence Intervals for the RUL

The bounds of the confidence interval at level (1–α) are obtained, as functions of time, by application of (5.25) and (5.26). The inverse function g^{-1}, which occurs in those equations, is readily derived from (5.30) as follows:

$$g^{-1}(x) = \eta \left[\frac{k-1}{k} \log \left(1 - \frac{k}{\mu} x \right) \right]^{\frac{1}{\beta}} \qquad (5.41)$$

defined for x ranging from 0 to ∞. The g^{-1} function is illustrated in Figure 5.3.

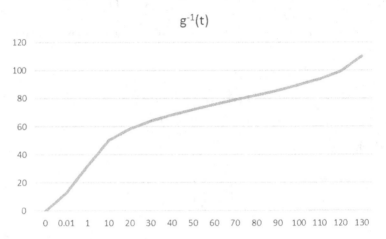

FIGURE 5.3   Inverse g^{-1} of the time transformation for 2P Weibull distribution (β = 5, η = 130h).

Using those results, the confidence interval for the RUL is obtained. It is illustrated in Figure 5.4 for two values of the shape parameter: β=2 and β=5, at an 80% confidence level (i.e. α=20%).

As expected, the interval is narrower for β=5.

Example

Let us consider an example, which will be further studied in Chapter 8 when we look at statistical estimation.

Light-emitting diodes (LEDs) tend to degrade with time, which results in a decrease in luminous flux. When that flux reaches a certain threshold (a percentage of the specified flux), the LED no longer performs according to specifications: thus, by the definition of the concept of failure (Chapter 2), it fails.

The speed of degradation will depend on the stresses to which the LEDs is subjected, essentially electrical current and temperature.

Assume for instance that the time to failure of a population of LEDs (with failure defined as above) is characterized by a two-parameter Weibull distribution

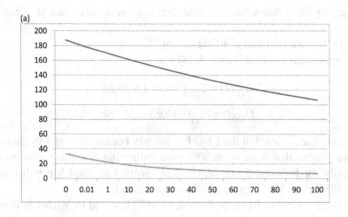

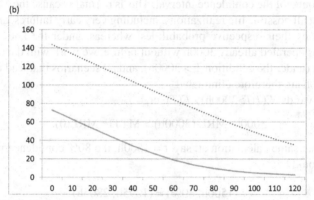

FIGURE 5.4 Confidence interval at 80% for RUL in the Weibull distribution, for two values of the shape parameter: (a) $\beta = 2$ and (b) $\beta = 5$.

with a scale parameter $\eta = 3347h$ and a shape parameter $\beta = 3.58$ (those are values which will be estimated from field data in Chapter 8).

Given that an LED has been operating according to specifications for 3000h, we would like to predict its RUL, with an 80% confidence interval. In other words, we would like to estimate how much longer the LED will keep operating correctly, beyond 3000h.

We will therefore apply (5.25) and (5.26), which become (5.27) and (5.28) for $\alpha = 20\%$, with

$t = 3000h$.

From (5.31),

$$\mu = \eta\Gamma\left(1+\frac{1}{\beta}\right) = 3347h\,\Gamma\left(1+\frac{1}{3.58}\right) = 3016h.$$

From (5.33), the value of k is derived: $k=0.82$.

We can see from that value of k that degradation is rather fast (k is not too far from 1).

From (5.30), we find that g(3000h) = 3503h.

Then, application of (5.27) and (5.28) results in

$$L_{0.2}(t) = L_{0.2}(3000h) = 59.6h$$

$$U_{0.2}(t) = U_{0.2}(3000h) = 152.6h$$

which means that, given that the LED has already been operating satisfactorily for 3000h, we predict that there is an 80% probability that it will continue to operate satisfactorily for between 59.6h and 152.6h, so that its total life will have been between 3059h and 3152h (with an 80% probability).

Note that the a priori life expectation, or MTTF, equal to 3016h, is lower than the lower bound of the confidence interval. This is normal because the MTTF takes into account all possible life realizations, including very early failures, and weights them down by their respective probabilities; whereas, under the knowledge that the LED has operated already 3000h without failure, we work with the conditional distribution (under the condition T > 3000h), and all scenarios that lead to a failure before 3000h are therefore eliminated.

In other words: $E(T|T>3000h) > E(T)$, or,

3000h + MRL (3000h) > MTTF = MRL(0).

If we did that same calculation at, say, t = 1500h, the 80% confidence interval for the RUL would be:

$$I(1500h, 80\%) = (1095h; 2803h)$$

which means that, given that the LED had been operating satisfactorily for 1500h, we would predict its total life to be between 2595h (= 1500h + 1095h) and 4303h (= 1500h + 2803h).

With such data, the confidence interval contains the a priori expectation 3016h. It is also much wider than with the previous example.

But the knowledge that the LED has operated correctly during 1500h is much less informative than in the previous case (where it had been seen to operate correctly for 3000h) and therefore, it teaches us much less in addition to what we knew already, i.e., the a priori probability distribution of the time to failure.

5.5   APPLICATION TO GAMMA DISTRIBUTION

The reliability function for the gamma distribution is given (Birolini 2017) as follows, in terms of a scale parameter λ and a shape parameter β:

$$R(t) = 1 - \frac{1}{\Gamma(\beta)} \int_0^{\lambda t} x^{\beta-1} e^{-x} dx. \tag{5.42}$$

The scale parameter has the dimension of the reciprocal of a time, and the shape parameter is dimensionless.

Expectation $E(T)$ and variance σ^2 can be expressed in terms of those two parameters:

$$E(T) = \frac{\beta}{\lambda} \qquad (5.43)$$

$$\sigma^2 = \frac{\beta}{\lambda^2}. \qquad (5.44)$$

Application of Theorem 5.1 leads to the following expression for the $g(.)$ function:

$$g(t) = \frac{\mu}{k}\left[1 - \left(1 - \frac{1}{\Gamma(\beta)}\int_0^{\lambda t} x^{\beta-1} e^{-x} \, dx\right)^{\frac{k}{1-k}}\right]. \qquad (5.45)$$

This expression contains the two parameters μ and k of the corresponding distribution with MRL linear in time into which the gamma distribution is transformed by $g(.)$. They can be derived in terms of the two parameters of the gamma distribution, by means of (5.4) and (5.8).

There follows:

$$\mu = \frac{\beta}{\lambda} \qquad (5.46)$$

$$k = \frac{\beta-1}{\beta+1}. \qquad (5.47)$$

It is verified that, as expected, the limiting case $k = 0$ corresponds to the exponential distribution ($\beta = 1$), and the limiting case $k = 1$ corresponds to the Dirac distribution and to $\beta \to \infty$.

This is consistent with the fact that, for the gamma distribution, the coefficient of variation CV vanishes asymptotically as $\beta \to \infty$, since

$$CV \equiv \frac{\sigma}{E(T)} = \frac{1}{\sqrt{\beta}}. \qquad (5.48)$$

Those results are summarized in Table 5.1.

The $g(.)$ function is plotted against time in Figure 5.5.

The expression in (5.45) contains the lower (regularized) incomplete gamma function. A short Python program to compute $g(t)$ is provided in Table 5.2. It takes β, λ, and t as inputs, and returns $g(t)$ as its output.

TABLE 5.1
Special Values of Shape Parameter β

	β	k	CV
Exponential	1	0	1
	3	1/2	$1/\sqrt{3}$
Dirac	∞	1	0

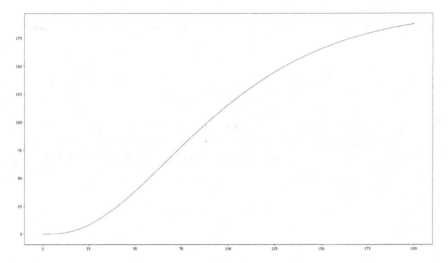

FIGURE 5.5 Time transformation $g(.)$ leading to the gamma distribution ($\beta = 3$, $\lambda = 0.01h^{-1}$).

In order to build confidence intervals, one needs the inverse function g^{-1}. In the case of the gamma function, it is not available in closed form, but it can be expressed in terms of the distribution quantiles, i.e. via the inverse R^{-1} of the reliability function:

$$g^{-1}(x) = R^{-1}\left[\left(1 - \frac{k}{\mu}x\right)^{\frac{1-k}{k}}\right] \qquad (5.49)$$

defined for $0 \le x < \dfrac{\mu}{k}$.

The quantiles of the gamma distribution are tabulated and can be obtained in terms of the inverse of the incomplete gamma function.

We can verify, on the example of the gamma distribution, the assertion of Section 5.2 about the higher-order moments.

TABLE 5.2
Python Program for $g(.)$ Transformation
Associated with Gamma Distribution

```
#Incomplete gamma function
import mpmath
from mpmath import gammainc
a = input ("enter shape parameter:")
lambd= input ("enter scale parameter:")
t = input("enter time:")
z = float(lambd)*float(t)
def gammainc_low(a,z):
    return gammainc(a, 0, z, regularized=True)
w = gammainc_low(a,z)
x = 1-w
from mpmath import *
u = fadd(a,1)
u1=fsub(a,1)
v = fdiv(u1,2)
y = power(x,v)
r = 1-y
#obtain mu, k and g
gn = fmul(a,u)
gd = float(lambd)*float(fsub(a,1))
g1 = fdiv(gn,gd)
g = g1*r
mu = fdiv(a, lambd)
print ("mu equals", mu)
print ("k equals", fdiv(u1,u))
print ("mu/k equals", g1)
print ("regularised gamma equals", w)
print ("multiplicand of mu/k equals", r)
print ("result g equals:", g)
```

For the gamma distribution, the nth order centered moment is given by

$$\alpha_n = \frac{\Gamma(\beta+n)}{\lambda^n \Gamma(\beta)} \qquad (5.50)$$

while, in the transformed distribution, the corresponding moment α'_n is given by (5.15).

For $n = 1$ and $n = 2$, $\alpha_n = \alpha'_n$:

Indeed, for $n = 1$, $\alpha_1 = \alpha'_1 = \dfrac{\beta}{\lambda}$ and, for $n = 2$, $\alpha_2 = \alpha'_2 = \beta\dfrac{(\beta+1)}{\lambda^2}$ – using (5.15) and (5.50).

But, for $n = 3$, $\alpha_3 \neq \alpha'_3$.

For instance, if $k = \frac{1}{2}$, then $\beta = 3$ from (5.47), and, from (5.50), $\alpha_3 = \frac{\Gamma(\beta+3)}{\Gamma(\beta)\lambda^3} = \frac{\Gamma(6)}{\Gamma(3)\lambda^3} = \frac{5!}{2!\lambda^3} = \frac{60}{\lambda^3}$, whereas, from (5.15), $\alpha'_3 = 3!(2\mu)^3 \frac{\Gamma(2)}{\Gamma(2+3)} = \frac{3!}{4!}(2\mu)^3$; and, from (5.46), since $\beta = 3$, $\mu = \frac{3}{\lambda}$,

therefore $\alpha'_3 = \frac{1}{4}\frac{2^3 3^3}{\lambda^3} = \frac{54}{\lambda^3} \neq \alpha_3$

Erlang Distribution

A sum of n independent, identically distributed exponential random variables, with failure rate λ, has a gamma distribution with scale parameter λ and shape parameter $\beta = n$. This is called an Erlang distribution.

This property is easily proved by means of the moment-generating functions: the MGF of an exponential with rate λ is

$$\Phi(s) = E\left(e^{sT}\right) = \frac{\lambda}{\lambda - s}.$$

The moment-generating function of the sum is the product of the moment-generating functions, which is equal to $\left(\frac{\lambda}{\lambda - s}\right)^n$, that is, the moment-generating function of the gamma distribution with scale parameter λ and shape parameter n, or Erlang (λ, n).

The Erlang distribution can be seen as depicting a sequence of n successive shocks: the first shock (after the first random time, exponentially distributed) causes a minor degradation of the asset and brings it to a mildly degraded state, where it still performs its function. The second shock (occurring after a second exponentially distributed time) further degrades it, and so on; after the nth shock, the function is lost, i.e., by definition, a failure takes place. Thus on average, the failure takes place after a time $\mu = \frac{n}{\lambda}$, i.e., n times the MTTF of one of the exponential random variables.

Example

Let us consider the same example as for the Weibull distribution: the LED.

Assume now the time-to-failure follows a gamma distribution, but with the same parameter k as before, i.e. $k = 0.82$. In the gamma distribution, from (5.47),

$$\beta = \frac{1+k}{1-k}.$$

Therefore, in this example, $\beta = 10.11$.

Also, we take the same value for the mean as in the Weibull example: $\mu = 3016h$, hence, from (5.46), $\lambda = \dfrac{\beta}{\mu} = 0.003352h^{-1}$.

The g transformation, from (5.45), gives $g(3000h) = 3566h$. (Using the Python program of Table 5.2.) Recall that, in the Weibull example, we had found $g(3000h) = 3503h$.

Thus, for the time value ($t = 3000h$), the gamma distribution accelerates the age a little more than the Weibull distribution. The effect is minor though, less than 2%.

5.6 APPLICATION TO THE LOGNORMAL DISTRIBUTION

As a reminder (Birolini 2017), a positive random variable has a lognormal distribution if its natural logarithm has a normal (Gaussian) distribution.

Thus, if τ denotes a lognormally distributed random variable, $\ln(\tau)$ is normally distributed.

If σ^2 denotes the variance of $\ln(\tau)$, the cumulative distribution function of τ can be expressed as:

$$F(t) = \phi\left(\frac{\ln(\lambda t)}{\sigma}\right), \tag{5.50}$$

where $\phi(.)$ denotes the cumulative distribution function of the normal standard deviate (i.e. normal with mean 0 and variance 1), and λ is a scale parameter whose dimension is the inverse of a time.
Equivalently

$$R(t) = 1 - F(t) = 1 - \frac{1}{\sqrt{2\pi}} \int_{-\infty}^{\frac{\ln(\lambda t)}{\sigma}} e^{-\frac{x^2}{2}} dx. \tag{5.51}$$

Then the expectation and variance of τ can be expressed as follows:

$$E(\tau) = \frac{e^{\frac{\sigma^2}{2}}}{\lambda} \tag{5.52}$$

$$\mathrm{Var}(\tau) = \frac{e^{2\sigma^2} - e^{\sigma^2}}{\lambda^2}. \tag{5.53}$$

One can then apply Theorem 5.1 to derive the function $g(t)$. The parameters μ and k are determined as usual, from (5.4) and (5.8):

$$\mu = \frac{e^{\frac{\sigma^2}{2}}}{\lambda} \tag{5.54}$$

$$k = 2e^{-\sigma^2} - 1. \tag{5.55}$$

As can be seen, k is a decreasing function of σ^2, and limiting cases correspond to:

$k = 1$ for $\sigma \to 0$ (degenerate deterministic case)

$k = 0$ for $\sigma = \sqrt{\ln(2)} \approx 0.83$.

The transformation $g(.)$ is then expressed as:

$$g(t) = \frac{\mu}{k} \left\{ 1 - \left[1 - \phi\left(\frac{\ln(\lambda t)}{\sigma} \right) \right]^{\frac{k}{1-k}} \right\} \tag{5.56}$$

with μ and k given by (5.54) and (5.55), respectively.

For $\sigma > \sqrt{\ln(2)}$, which corresponds to a coefficient of variation greater than 1 for τ, a negative value of k is obtained.

5.7 APPLICATION TO THE PARETO DISTRIBUTION

The Pareto distribution (Type 1) is characterized by the following reliability function:

$$R(t) = \begin{cases} \left(\dfrac{b}{t}\right)^{\alpha} & t > b \\ 1 & t \leq b \end{cases} \tag{5.57}$$

with two positive parameters: b and α.

It can be transformed into a distribution with an MRL linear function of time only within a certain range of those parameters.

For $\alpha > 1$, the expectation is given by

$$\mu = \frac{\alpha}{\alpha - 1} b \tag{5.58}$$

and, for $\alpha > 2$, the variance is expressed as

$$\sigma^2 = \frac{\alpha b^2}{(\alpha - 1)^2 (\alpha - 2)}. \tag{5.59}$$

There follows from (5.8) that

$$k = \frac{\alpha(\alpha - 2) - 1}{\alpha(\alpha - 2) + 1}. \tag{5.60}$$

Limiting cases are:

i. for $\alpha \to \infty$, lim $k = 1$. This case corresponds to the Dirac distribution at $t = b$, as indeed (5.57) shows that the reliability function then converges to a step function with the step at $t = b$.

ii. for $\alpha = \sqrt{2} + 1 \approx 2.414$, $k = 0$ which corresponds to a coefficient of variation equal to 1.

For smaller values of the parameter α, the corresponding coefficient k would be negative.

For the range $\alpha \geq 2.414$, the following transformation $g(.)$ converts the Pareto distribution into a distribution with MRL linear in time:

$$g(t) = \frac{\alpha}{(\alpha - 1)} \frac{\left[\alpha(\alpha - 2) + 1\right]}{\left[\alpha(\alpha - 2) - 1\right]} b \left[1 - \left(\frac{b}{t}\right)^{\frac{\alpha[\alpha(\alpha-2)-1]}{2}}\right] \quad (5.61)$$

for $t \geq b$ and 0 for $t \leq b$.

This function is illustrated in Figure 5.6. It is a concave increasing function. The higher the value of the parameter α, the closer $g(.)$ approaches a step function, with the step at $t=b$, corresponding to a deterministic life equal to b.

Note that, unlike the g transformations for the Weibull and gamma distributions, which are S-shaped (first convex, then concave, with an inflection point), the g transformation for the Pareto distribution is concave.

The general behavior of the g transformation will be studied in detail in Chapter 6.

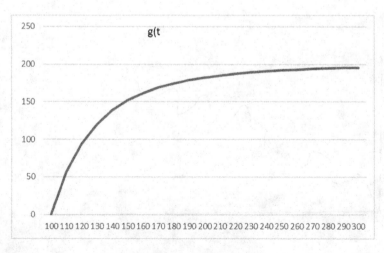

FIGURE 5.6 Time transformation for Pareto distribution.

5.8 APPLICATION TO CONTINUOUS DEGRADATION PROCESSES

5.8.1 PROBLEM STATEMENT

A number of degradation mechanisms can be modeled by the time evolution of a variable that measures the severity of the degradation: for instance, the size of a crack, the amount of dust accumulation in a filter, the magnitude of the drift in an electrical circuit parameter, and so on.

Since there is some uncertainty in the corresponding values, the process is not deterministic, and is therefore best modeled by a stochastic process.

When the degradation reaches a critical threshold, the function is lost, i.e. a failure occurs.

Therefore, the problem of calculating the RUL is equivalent to that of calculating the time necessary for the stochastic process to cross the critical threshold, the so-called first hitting time (Karlin & Taylor 1975), as illustrated in Figure 5.7. The probability distribution of the RUL can then be derived from that of the stochastic process that models the degradation evolution over time.

The stochastic processes which can be used in order to model the evolution of degradations are quite numerous. We shall study here two processes, which have properties that make them amenable to such a description: the Wiener process and the Gamma process.

They share the following properties (Birolini 2017; Karlin & Taylor 1975, Huynh et al. 2019, Abdel-Hameed 2014) which define the class of "Lévy processes":

1. independent increments: the increments of the processes in disjoint time intervals are statistically independent random variables;

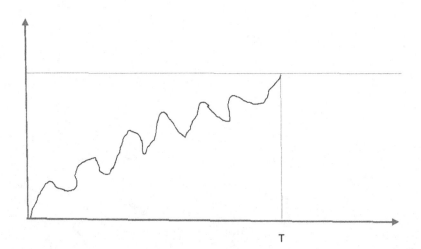

T

FIGURE 5.7   Concept of first hitting time.

mathematically, denoting X_t the random variable "process X at time t", this property means that, for any times $t_1 < t_2 < \ldots t_n$, the variables $(X_{t_2} - X_{t_1})$, $(X_{t_3} - X_{t_2})$, $\ldots (X_{t_n} - X_{n-1})$ are stochastically independent;

2. stationary increments: the probability distribution of the increment $(X_{t+h} - X_t)$ depends only on the length of time h.

The theory of Section 5.6 is now applied to the Wiener process and to the Gamma process.

The key difference between those two families is that the Gamma process is monotonic (degradations can only increase with time), while the Wiener process is not.

Examples of physical situations which those models can describe are numerous. Let us mention a few:

- crack growth propagation;
- rail track defect propagation;
- degradation of pipelines due to corrosion;
- degradation of power electronics components (e.g. MOSFET-SiC);
- degradation of batteries.

5.8.2 RUL FOR WIENER PROCESS WITH DRIFT

In general, the Wiener process *with drift* can be described by the property that the increment $(X_{s+t} - X_s)$, over a time period $(s, s + t)$, has a normal (Gaussian) distribution, with mean equal to:

$$E(X_{s+t} - X_s) = vt \tag{5.62}$$

and variance equal to:

$$\mathrm{Var}(X_{s+t} - X_s) = \sigma^2 t. \tag{5.63}$$

Thus, both the mean and the variance are linear functions of time.

(Note therefore that, if X_t is measured in m, then v is measured in $m \cdot h^{-1}$, and σ^2 is measured in $m^2 \cdot h^{-1}$.)

If X_s represents the magnitude of a degradation (for instance, the size of a crack), at time s, then vt is the average degradation increase, i.e. the drift, over a time period of duration t; and the parameter v is the average drift per unit of time.

Thus the Wiener process captures the fact there is a linear increasing trend in degradation, with random fluctuations around the mean (see Figure 5.8 for an illustration).

Thus the cumulative distribution function of the increment $(X_{s+t} - X_s)$ is expressed as:

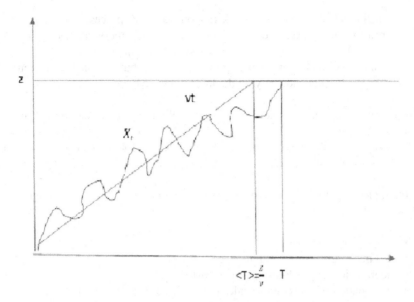

FIGURE 5.8 First hitting time T for Wiener process with drift parameter v.

$$P\left(X_{s+t} - X_s < x\right) = \frac{1}{\sigma\sqrt{2\pi t}} \int_{-\infty}^{x} e^{-\frac{(u-vt)^2}{2\sigma^2 t}} \, du. \tag{5.64}$$

This is the probability that, over a time period of duration t, the incremental degradation (e.g. the increase in crack size) does not exceed x.

If z denotes the critical threshold, the first hitting time T is the time needed for the degradation to reach the level z, assuming no degradation at initial time $t=0$.

By definition,

$$T = \inf\left\{t : X_t \geq z\right\}, \tag{5.65}$$

where T has an inverse Gaussian distribution (Bagdonavicius & Nikulin 2002): the probability distribution of T is described by the probability density $f(t)$:

$$f(t) = \frac{z}{\sigma\sqrt{(2\pi t^3)}} e^{-\frac{(z-vt)^2}{2\sigma^2}}. \tag{5.65}$$

The expectation and variance of T can be expressed in terms of the Wiener process distribution.

It is convenient to use the following notation (Folks & Chhikara 1978):

$$\mu = \frac{z}{v} \tag{5.66}$$

$$\lambda = \left(\frac{z}{\sigma}\right)^2. \tag{5.67}$$

Both μ and λ have the dimension of a time.
   Then

$$E(T) = \mu \tag{5.68}$$

$$\mathrm{Var}(T) = \frac{\mu^3}{\lambda}. \tag{5.69}$$

Hence the coefficient of variation of T is given by

$$cv = \sqrt{\frac{\mu}{\lambda}}. \tag{5.70}$$

According to Theorem 5.1, the parameter k is obtained as follows:

$$k = \frac{1 - cv^2}{1 + cv^2} = \frac{\lambda - \mu}{\lambda + \mu}. \tag{5.71}$$

In order for the distribution of the first hitting time T to be able to be converted into a distribution with MRL linear in time according to some time transformation $g(.)$, the condition $k \geq 0$ is required. According to (5.71), this condition amounts to the coefficient of variation of T not exceeding unity:

$$\lambda \geq \mu \tag{5.72}$$

which, given (5.66) and (5.67), is equivalent to:

$$\frac{\sigma^2}{v} \leq z. \tag{5.73}$$

The interpretation of this constraint is the following.
   Denoting by t^* the *average* time needed to reach the degradation level z,

$$t^* = \frac{z}{v} \tag{5.74}$$

the standard deviation of the degradation at that time t^* is equal to:

$$\mathrm{Stdev}(X_{t^*}) = \sigma\sqrt{t^*} = \sigma\sqrt{\frac{z}{v}}.$$

The constraint (5.73) states that the standard deviation at time t^* must not exceed the average degradation at that time.

Thus a nonlinear time transformation can map the random variable T into a random variable with MRL linear in time only if the average fluctuations, characterized by σ, are not larger than the average drift, characterized by v. Otherwise stated, for such a transformation to be possible, noise should not exceed signal, i.e. the signal-to-noise ratio should not be lower than 1.

Since the Wiener process is not monotonically increasing, for small degradation levels (i.e. values of z which do not satisfy (5.73)), fluctuations, characterized by σ, could be more significant that the drift, characterized by v.

When constraint (5.73) is met, the theory of Section 5.6 can be applied to derive the $g(.)$ transformation., from the reliability function.

The cumulative distribution function of the inverse Gaussian distribution can be expressed explicitly in terms of the standard normal deviate, $\Phi(t)$, see e.g. (Folks & Chhikara 1978).

Limiting Cases

1. $k = 0$ corresponds to the limiting case $z = \dfrac{\sigma^2}{v}$.
2. $k = 1$ corresponds to $\mu = 0$, which is possible, either if $z = 0$ (i.e. the degradation threshold is immediately reached) or $v \to \infty$ (the drift is infinitely fast). In both cases, those limiting conditions mean a sudden failure, a Dirac function (consistent with $k = 1$).

5.8.3   RUL FOR GAMMA PROCESS

The Gamma process is another process with stationary, independent increments which, unlike the Wiener process, is monotonically increasing. Therefore it is quite adequate to model cumulative degradation, and has been used extensively for that purpose (Van Noortwijk 2009).

Mathematically it is characterized as follows.

The Gamma process X_t has stationary independent increments $X_{t+s} - X_s$ which are distributed according to a gamma distribution of shape parameter β (t) and scale parameter λ (see Section 5.5), namely:

$$f(t) = \frac{\lambda}{\Gamma(\beta(t))}(\lambda t)^{\beta(t)-1} e^{-\lambda t}. \tag{5.75}$$

For some shape function β(t).

In the most common case, the shape parameter function β(t) is linear:

$$\beta(t) = \beta t \tag{5.76}$$

(with β measured in h^{-1}), and the probability density function (5.65) becomes:

$$f(t) = \frac{\lambda}{\Gamma(\beta t)}(\lambda t)^{\beta t-1} e^{-\lambda t}.$$ (5.77)

The increment over a period of duration t, $X_{t+s} - X_s$ therefore has a mean equal to

$$E(X_{t+s} - X_s) = \frac{\beta t}{\lambda}$$ (5.78)

and a variance equal to

$$\text{Var}(X_{t+s} - X_s) = \frac{\beta t}{\lambda^2}.$$ (5.79)

Therefore the variance-over-mean ratio remains constant with time (equal to $1/\lambda$). The coefficient of variation, on the contrary, decreases in time, since

$$CV = \frac{E}{\sqrt{\text{Var}}} = \frac{1}{\sqrt{\beta t}}.$$ (5.80)

The assumption (5.76) expresses the fact that the average deterioration is a linear function of time (5.78); see e.g. (Van Noortwijk 2009).

If the critical threshold is denoted z, then the RUL is the first hitting time of a threshold z (a degradation of magnitude z) from level 0.

Its reliability function is given by (Van Noortwijk 2009):
$$R(t;z) = P[\text{RUL} > t] = P[FHT(z) > t] = P[X_t - X_0 < z]$$

$$R(t;z) = \frac{\int_0^{\lambda z}(x)^{\beta t-1} e^{-x}dx}{\int_0^{\infty}(x)^{\beta t-1} e^{-x}dx} = \frac{\Gamma(\beta t, \lambda z)}{\Gamma(\beta t)},$$ (5.81)

where $\Gamma(\beta, y)$ denotes the lower incomplete gamma function:

$$\Gamma(\beta, y) = \int_0^y x^{\beta-1} e^{-x}dx.$$ (5.82)

From this expression it is not possible to derive in closed form the mean and variance of the RUL, therefore the parameters μ and k. They can however be estimated statistically.

The density function can be obtained by differentiation. It can be approximated by the Birnbaum–Saunders distribution (Van Noortwijk 2009) when the coefficient of variation is small.

5.9   EXERCISES

1. Derive a formula for the failure rate of the Weibull distribution. Calculate the failure rate in the example of Section 5.4.2, at $t = 3000$h.

2. Calculate α_3, the third-order centered moment for the Weibull distribution with shape parameter equal to 2 (also called the Rayleigh distribution), and show it is not equal to α_3', the third-order centered moment of the transformed distribution (by the mapping $g(t)$)

3. In the LED example of Section 5.4.2, calculate the bounds of the 80% confidence interval of RUL(t), for t ranging from 0 to 5 000h (by increments of 500h). What do you notice?

4. Same as Exercise 2 but for $\beta = 2$. Compare the results with those of Exercise 3.

5. Consider a system which is subjected to repeated, independent shocks, with, on average, one shock every 24h. It is considered that the system fails after undergoing three successive shocks.
    a. Calculate the MTTF.
    b. What is the probability that the system operates without failure for at least one week?
    c. Calculate the standard deviation of the TTF.

6. In a nuclear power plant, a crack has been found in a vessel. The crack size is growing at an average rate of 0.5 mm per year, with a variance of 0.01 mm²/year. A crack size of 2 mm is considered a safety hazard and would cause the plant operation to be halted.
    a. What is the average time before the crack reaches the critical size?
    b. What is the probability that the plant will have to be stopped before 4 years?
    c. Can the time-to-failure be transformed into a distribution with MRL linear in time? If so, please write the g transformation and calculate the k coefficient. (Hint: consider a Wiener process with drift.)

7. Two designs for a track circuit (a circuit that enables train presence detection on a portion of rail track) are compared. Reliability studies conclude that the first design results in a constant failure rate $\lambda=2.10^{-5}$/h; while the second design leads to a two-parameter Weibull distribution for the time to failure, with same mean as in the first design but a shape factor $\beta=3$.

    Suppose the track circuit has been operating failure free for 40 000h. Determine the 80% confidence interval for RUL(t) at $t= 40$ 000h, for each of the two designs.

    As a maintainer, which design would you prefer? Why?

BIBLIOGRAPHY

Abdel-Hameed, M., *Levy Processes and Their Applications in Reliability and Storage.* Springer Briefs in Statistics, Switzerland, Springer, 2014.

Bagdonavicius, V., & Nikulin, M.S., *Accelerated Life Models* Boca Raton, FL, Chapman & Hall, CRC, 2002.

Birolini, A., *Reliability Engineering, Theory & Practice*. Springer, New York, 8th Edition, 2017.

Folks, J.L., & Chhikara, R.S, "The Inverse Gaussian Distribution and Its Statistical Application—A Review", *Journal of Royal Statistical Society B*, 40(3), 263–289, 1978.

Gumbel, E.J., *Statistics of Extremes*. New York, Columbia University Press, 1958.

Huynh, K., Grall, A., Bérenguer, C. A Parametric Predictive Maintenance Decision-Making Framework Considering Improved System Health Prognosis Precision", *IEEE Transactions on Reliability*, 2019, 68(1), 375–396. 10.1109/TR.2018.2829771. hal-01887

Karlin, S., & Taylor, H.M., *A First Course on Stochastic Processes*. London, Academic Press, 1975.

Nachlas, J., *Reliability Engineering: Probabilistic Models and Maintenance Methods*. Boca Raton, FL, CRC Press, Taylor & Francis, 2005.

Van Noortwijk, J.M., "A Survey of the Application of Gamma Processes in Maintenance", *Reliability & System Safety*, 94, 2–21, 2009.

Weibull, W., "A Statistical Theory of the Strength of Materials", *Proceeding of the Swedish Royal Institute of Engineering Research*, p. 153, 1939.

6 Properties of the *dg* Metric

6.1 INTRODUCTION

The $g(.)$ transformation introduced in Section 5.1 (Theorem 5.1) defines a metric.

More specifically, its derivative $\dfrac{dg}{dt}$ measures the rate of change of transformed time with respect to initial time that is necessary to make the mean residual life (MRL) linear.

The link between transformed time t' and initial time t is obviously:

$$dt' = \frac{dg}{dt}\, dt. \tag{6.1}$$

Equation (6.1) shows that, if $\dfrac{dg}{dt}$ increases with time t, the same increment in transformed time dt' is achieved with a smaller increment dt in initial time. Since the increment dt' is linked linearly to the MRL V', an increasing $\dfrac{dg}{dt}$ function describes aging that accelerates with time, in the sense that, as time (the initial one) progresses, the asset ages more quickly, i.e. its remaining useful life (RUL) is reduced more quickly on average, or its average RUL loss rate increases.

Conversely, if $\dfrac{dg}{dt}$ decreases with time, the same increment in transformed time dt' is achieved with a larger increment dt in initial time. The MRL will then decrease more and more slowly, i.e. on average, the loss of RUL per unit of time decreases.

In view of those remarks, it is of interest to study the time derivative of g, and actually also its second derivative.

DOI: 10.1201/9781003250685-6

6.2 DERIVATIVE OF g(T)

From the definition (5.6) for $g(.)$, there follows:

$$\frac{dg}{dt} = -\frac{\mu}{1-k} R(t)^{\frac{k}{1-k}-1} \frac{dR}{dt} = -\frac{\mu}{1-k} R(t)^{\frac{2k-1}{1-k}} \frac{dR}{dt} \tag{6.2}$$

And, since by definition of the failure rate $\lambda(t)$,

$$\lambda(t) = -\frac{1}{R}\frac{dR}{dt}, \tag{6.3}$$

(6.2) can be written as

$$\frac{dg}{dt} = \frac{\mu}{1-k} \lambda(t) R(t)^{1+\frac{2k-1}{1-k}} \tag{6.4}$$

or equivalently,

$$\frac{dg}{dt} = \frac{\mu}{1-k} \lambda(t) R(t)^{\frac{k}{1-k}} \tag{6.5}$$

First, (6.5) shows that $\frac{dg}{dt} \geq 0$, i.e. $g(t)$ is a monotonic, non-decreasing function.

From (6.5) and since $\lim_{t \to \infty} R(t) = 0$, it is seen that the $g(t)$ curve has a horizontal asymptote:

$$\lim_{t \to \infty} \frac{dg}{dt} = 0.$$

This property expresses the fact that the marginal "time acceleration" is asymptotically zero. Recall that $g(t)$ converges to $\frac{\mu}{k}$ asymptotically.

Also, (6.5) shows that

$$\frac{dg}{dt}(0) = \frac{\mu}{1-k} \lambda(0) \geq 0. \tag{6.6}$$

In the special case where $k=0$ (exponential distribution), (6.6) reduces to

$$\frac{dg}{dt}(0) = \lambda\mu = 1$$

which is to be expected since, in that case, $g(t) = t$.

Based on those observations, there follows that there are two possible behaviors for the function $g(t)$:

- either it is concave (as in the case of the Pareto distribution);
- or it is first convex, for small values of t, and then concave for larger values of t; which means that it has at least one point of inflection (the Weibull and gamma distributions fall into that category, with one point of inflection).

Analysis of the second derivative (Section 6.3) will confirm and refine those conclusions.

6.3 DIFFERENTIAL EQUATION FOR MRL

There is a differential relation which links the metric dg to the average RUL loss, i.e. $\dfrac{dV}{dt}$. It is an immediate consequence of the definition of the transformation $g()$.

Indeed, recall (Section 5.2) that

$$R(t) = R'(t') \qquad (6.7)$$

i.e. reliability is invariant under the time transformation defined by $g()$. Consequently,

as $t' = g(t)$,

$$\frac{dR(t)}{dt} = \frac{dR'(t')}{dt} = \frac{dR'}{dt'} \frac{dg}{dt}. \qquad (6.8)$$

Introducing the failure rates, $\lambda(t)$ and $\lambda'(t')$, and taking (6.7) and (6.8) into account,

$$\lambda(t) = -\frac{\dfrac{dR}{dt}}{R(t)} = -\frac{\dfrac{dR'}{dt'}}{R'(t')} \frac{dg}{dt} = \lambda'(t') \frac{dg}{dt} \qquad (6.9)$$

which shows that one interpretation of $\dfrac{dg}{dt}$ is the ratio of the failure rates expressed in the initial and the transformed times:

$$\frac{dg}{dt} = \frac{\lambda(t)}{\lambda'(t')}. \qquad (6.10)$$

Now, in the transformed time, the expression of the failure rate is known to be:

$$\lambda'(t') = \frac{1-k}{\mu - kt'} \qquad (6.11)$$

since it corresponds to a time-to-failure distribution with MRL linear in time (Chapter 3).

The fundamental relation which links MRL with failure rate (Chapter 2) can be written as:

$$\frac{dV}{dt} = \lambda(t)V(t) - 1 = \lambda'(t')\frac{dg}{dt} V(t) - 1.$$

Or, in view of (6.11),

$$\frac{dV}{dt} = \frac{1-k}{\mu - kg(t)} \frac{dg}{dt} V(t) - 1. \tag{6.12}$$

If $g(t)$ is known, (6.12) can be seen as a differential equation in $V(t)$. It can be solved with the initial condition

$$V(0) = \text{MTTF} = \mu.$$

Note that it is a linear differential equation (with variable coefficients).

Conversely, (6.12) can be written as:

$$\frac{dg}{dt} = \frac{1 + \dfrac{dV}{dt}}{(1-k)} \frac{(\mu - kg(t))}{V(t)}. \tag{6.13}$$

If $V(t)$ is known, (6.13) can be seen as a linear differential equation for $g(t)$.

The initial condition is: $g(0) = 0$.

It is easily verified that, if $g(.)$ is the identity function, i.e. $g(t) = t$, then $V(t)$ is linear:

$$V(t) = \mu - kt.$$

More generally, invertible nondecreasing differentiable functions $g(t)$ that map the nonnegative half real line onto a finite interval $(0,a)$ can be used to define lifetime distributions.

6.3.1 Example: The Rayleigh Distribution

In Figure 6.1, the functions $V(t), \dfrac{dV}{dt}, g(t)$ and $\dfrac{dg}{dt}$ are plotted for the Rayleigh distribution, which is the Weibull distribution with a shape factor equal to 2. For that distribution, $V(t)$ can be expressed explicitly (Lai 2004) as follows:

$$V(t) = \sqrt{2\pi}[1 - \phi(t)].e^{\frac{t^2}{2}} \tag{6.14}$$

where $\phi(t)$ denotes the cumulative distribution function of the standard normal deviate (expectation 0 and variance 1); the scale variable η has been taken equal to $\sqrt{2}$ to facilitate calculations.

The reliability function is expressed as:

$$R(t) = e^{-\frac{t^2}{2}}. \tag{6.15}$$

From Chapter 5, (5.31) and (5.33),

$$\mu = \eta\Gamma\left(1+\frac{1}{\beta}\right) = \sqrt{2}\Gamma\left(\frac{3}{2}\right) = \sqrt{2}\,\frac{\Gamma\left(\frac{1}{2}\right)}{2} = \sqrt{2}\,\frac{\sqrt{\pi}}{2} = \frac{\sqrt{2\pi}}{2} = V(0) \tag{6.16}$$

and

$$k = \frac{2\Gamma^2\left(1+\frac{1}{\beta}\right)}{\Gamma\left(1+\frac{2}{\beta}\right)} - 1 = \frac{\pi}{2} - 1 \approx 0.57. \tag{6.17}$$

Therefore the function $g(.)$ is obtained explicitly in that case:

$$g(t) = \frac{\sqrt{2\pi}}{\pi-2}\left(1-e^{-\frac{(\pi-2)\frac{t^2}{2}}{4-\pi}}\right). \tag{6.18}$$

Figure 6.1 shows plots of $V(t)$, $g(t)$ and dg/dt in that example.

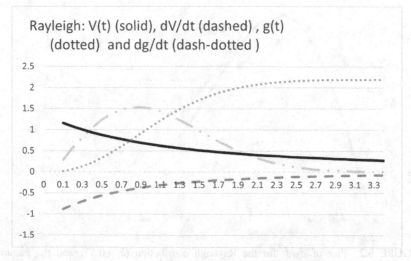

FIGURE 6.1 MRL Function and Metric (g and dg/dt) for Rayleigh Distribution.

6.3.2 OTHER EXAMPLE: THE GAMMA DISTRIBUTION

Let us apply (6.5) to the Gamma distribution (with reliability function $R(t)$ given by (5.32)).

First let us express the failure rate:

$$\lambda(t) = -\frac{\dfrac{dR}{dt}}{R(t)} = \frac{\lambda(\lambda t)^{\beta-1}}{\Gamma(\beta)R(t)}e^{-\lambda t}. \tag{6.19}$$

Then

$$\frac{dg}{dt} = \frac{\mu}{1-k}\frac{\lambda(\lambda t)^{\beta-1}}{\Gamma(\beta)R(t)}e^{-\lambda t}R(t)^{\frac{k}{1-k}} = \frac{\mu}{1-k}\frac{\lambda(\lambda t)^{\beta-1}}{\Gamma(\beta)}e^{-\lambda t}\,R(t)^{\frac{2k-1}{1-k}}. \tag{6.20}$$

In the special case $\beta = 3$, the corresponding value of k is $k = \dfrac{\beta-1}{\beta+1} = \dfrac{1}{2}$ (from (5.37)), and therefore:

$$\frac{dg}{dt} = 3\left(\lambda t\right)^2 e^{-\lambda t} \tag{6.21}$$

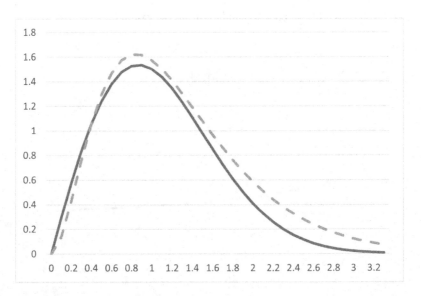

FIGURE 6.2 Plot of dg/dt for the Rayleigh distribution ($k \approx 0.57$) and the gamma distribution (shape factor 3, $k = 0.50$).

It is then possible to compare the metric $\dfrac{dg}{dt}$ for the Rayleigh distribution (which corresponds to $k \approx 0.57$) with the metric for the above Gamma distribution (with $k = 0.50$).

They happen to be very close, as illustrated in Figure 6.2.

From this example at least, it would seem that, beyond the particulars of the distribution, the k parameter is a determinant factor of the metric $\dfrac{dg}{dt}$.

From this analysis it appears that Weibull with shape factor 2 (Rayleigh) is very close to Gamma with shape factor 3.

6.4   SECOND DERIVATIVE OF g(T)

Let us now take a look at the second derivative of the transformation $g(.)$
By differentiating (6.5), the following is obtained:

$$\frac{d^2g}{dt^2} = \frac{\mu}{1-k}\left(\frac{d\lambda}{dt} - \frac{k}{1-k}\lambda^2\right)R(t)^{\frac{k}{1-k}} \tag{6.22}$$

The sign of the second derivative (left-hand side) is the same as that of the expression between parentheses in the right-hand side.

If the failure rate decreases with time, $\dfrac{d\lambda}{dt}$ is negative, the expression between parentheses is negative, and so is $\dfrac{d^2g}{dt^2}$: the g function is then concave.

In case of an increasing failure rate, $\dfrac{d\lambda}{dt}$ is positive, and there may be values of t for which the parenthesis vanishes, i.e. g has an inflection point.

Let us look for instance at the special cases of the Weibull and Gamma distributions.

6.4.1   WEIBULL DISTRIBUTION

The inflection point for $g(t)$ is the value t^* of t for which the expression between parentheses in (6.22) vanishes, i.e.

$$\frac{d\lambda}{dt} - \frac{k}{1-k}\lambda^2 = 0. \tag{6.23}$$

In the Weibull distribution case,

$$\lambda(t) = \frac{\beta}{\eta}\left(\frac{t}{\eta}\right)^{\beta-1}. \tag{6.24}$$

Therefore (6.23) becomes:

$$\frac{\beta}{\eta^2}(\beta-1)\left(\frac{t}{\eta}\right)^{\beta-2} - \frac{k}{1-k}\left(\frac{\beta}{\eta}\right)^2\left(\frac{t}{\eta}\right)^{2(\beta-1)} = 0 \qquad (6.25)$$

which, for $\beta > 1$, has the following unique solution t^*:

$$t^* = \eta\left(\frac{\beta-1}{\beta}\frac{1-k}{k}\right)^{\frac{1}{\beta}}. \qquad (6.26)$$

(Note that, in the exponential case, $\beta=1$, $k=0$, (6.25) is identically equal to 0).

In the example of Figure 6.2, $\beta = 2$ and $\eta = \sqrt{2}h$, then $k = \frac{\pi}{2}-1 \approx 0.57$. There follows:

$$t^* = \sqrt{2}\sqrt{\frac{4-\pi}{2(\pi-2)}} \approx 0.867h$$

$$\left(\text{and } \mu = \eta\Gamma\left(1+\frac{1}{\beta}\right) = \sqrt{2}\Gamma\left(\frac{3}{2}\right) = \frac{\sqrt{2}}{2}\Gamma\left(\frac{1}{2}\right) = \frac{\sqrt{2\pi}}{2} \approx 1.25h\right).$$

From (6.26) it is possible to infer the behavior of t^* as a function of the shape parameter.

As β increases to ∞ and therefore k converges to 1, Lim $t^* = \eta$. This limiting case corresponds to the Dirac distribution, for which the $g(.)$ transformation is a step function, with the step is located at the mean, which is η in the limit.

At the other extreme, $\beta=1$ or $k=0$ corresponds to the exponential function, and then t^* is indeterminate (no inflection point, $g(.)$ is linear).

6.4.2 GAMMA DISTRIBUTION

In the example considered ($\beta=3$), $\frac{dg}{dt}$ is given by (6.21). There follows that $\frac{d^2g}{dt^2}$ vanishes

$$\text{for } t = \frac{2}{\lambda} = t^*.$$

In the example, with the same mean μ as for the Rayleigh distribution,

$$\lambda = \frac{\beta}{\mu} = 3*\frac{2}{\sqrt{2\pi}} \approx 2.39h^{-1}.$$

Therefore

$t^* = (2/2.39)$ h $= 0.835h$, which is very close to the corresponding value for the Rayleigh distribution, as can also be seen in Figure 6.2.

6.5   UPPER BOUND FOR THE AVERAGE RUL LOSS RATE

The meaning of this inflection point t^* studied in Section 6.4 is that it represents a sort of turning point, after which aging (understood as loss rate of RUL) becomes slower on average—whereas, before time t^*, it kept increasing.

More precisely, it will now be shown that, for values of t higher than t^*, it is possible to derive an upper bound for the absolute value of the average RUL loss rate, i.e. $\left|\dfrac{dV}{dt}\right|$.

Theorem 6.1

In the case of an S-shaped transformation $g(t)$, and if the function $\dfrac{1}{\lambda(t)}$ is strictly convex; let t^* denote the inflection point; then, for $t > t^*$, the absolute value of the average RUL loss rate is bounded by the parameter k, i.e. the slope of the MRL in transformed time:

$$\left|\frac{dV}{dt}\right| \le k \text{ for } t > t^*.$$

Proof
By definition of t^*, there holds, for $t > t^*$,

$$\frac{d^2 g}{dt^2} \le 0$$

which, in view of the expression (6.22), is equivalent to:

$$\frac{d\lambda}{dt} - \frac{k}{1-k}\lambda^2 \le 0. \tag{6.27}$$

Let us now differentiate both sides of the "fundamental equation" (Chapter 2):

$$\frac{dV}{dt} = \lambda(t)V(t) - 1. \tag{6.28}$$

There follows:

$$\frac{d^2 V}{dt^2} = \frac{d\lambda}{dt}V + \lambda\frac{dV}{dt} = \frac{d\lambda}{dt}V + \lambda(\lambda V - 1) \tag{6.29}$$

or

$$\frac{d^2 V}{dt^2} = \left(\frac{d\lambda}{dt} + \lambda^2\right)V - \lambda. \tag{6.30}$$

In view of (6.27), (6.30) implies

$$\frac{d^2V}{dt^2} \leq \left(\frac{k}{1-k} \lambda^2 + \lambda^2 \right) V - \lambda = \lambda \left(\frac{\lambda V}{1-k} - 1 \right). \tag{6.31}$$

It is known (Banjevic 2009) that, if $\frac{1}{\lambda(t)}$ is strictly convex, then $V(t)$ is convex.

Therefore, if $\frac{1}{\lambda(t)}$ is strictly convex,

$$\frac{d^2V}{dt^2} \geq 0.$$

Then (6.31) implies:

$$\frac{\lambda V}{1-k} - 1 \geq 0 \tag{6.32}$$

Or

$$\lambda(t)V(t) \geq 1-k \quad \text{for} \quad t \geq t^*. \tag{6.33}$$

But also, according to (6.28),

$$\lambda V = \frac{dV}{dT} + 1. \tag{6.34}$$

Therefore (6.33) is equivalent to:

$$\frac{dV}{dT} + 1 \geq 1 - k.$$

Hence

$$\frac{dV}{dT} \geq -k \quad \text{for} \quad t > t^*$$

and, since $\frac{dV}{dT} \leq 0$, this is equivalent to $\left| \frac{dV}{dt} \right| \leq k$ qed.

Examples
Consider the Rayleigh distribution studied in Section 6.2. It was seen that $t^* = 0.867\text{h}$ and that $k = 0.57$.

Therefore one can say that, for $t > 0.867\text{h}$, $\left| \frac{dV}{dt} \right| \leq 0.57$.

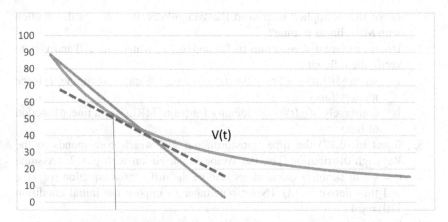

FIGURE 6.3 MRL of Rayleigh distribution (η = 100h) with slope (dashed line) less than $k = 0.57$ in absolute value for $t \geq t^* = 61\,h$. Solid line shows slope k. t* is indicated by vertical line.

Consider the Gamma distribution with scale factor of 3 and shape factor 2.39 h^{-1} of Section 6.2. It was seen that $t^* = 0.835h$ and $k = 0.50$.

One can therefore state that, for $t > 0.835h$, $\left|\dfrac{dV}{dt}\right| \leq 0.50$.

If now the exponential function is considered, or more generally a distribution with MRL linear in time, then, in (6.22), the left-hand side vanishes, and it is verified that so does the right-hand side, from the expression for λ, which is then (Chapter 3):

$$\lambda = \frac{1-k}{\mu - kt}. \tag{6.35}$$

In that limiting case,

$$\left|\frac{dV}{dt}\right| = k. \tag{6.36}$$

Figure 6.3 displays the plot of the MRL function for the Rayleigh distribution with η= 100h (shown in Chapter 2) with the tangent at $t= t^*$; which is seen to have a slope less steep than k (the straight line with slope k is also displayed).

6.6   EXERCISES

1. Consider the Weibull distribution with shape factor equal to 3 and scale factor equal to 100h. After a time $t= 50h$, what is the time acceleration

factor that is applied to convert the Weibull distribution to a distribution with MRL linear in time?

2. Use the differential equations (6.12) and (6.13), which link $g(t)$ and $V(t)$, to verify the following:

   a. when MRL(t) is a linear function of time, the transformation $g(t)$ is the identity function:

   b. conversely, if $g(t)$ is the identity function, MRL(t) is a linear function of time.

3. Inject in (6.12) the time transformation $g(t)$ which corresponds to the Rayleigh distribution, i.e., the Weibull distribution with $\beta = 2$. (Assume $\eta = \sqrt{2}$ to simplify calculations.) Solve the differential equation for $V(t)$, and thus derive (6.14). (Note: remember to impose the initial condition $V(0) = \mu$.)

4. Find the shape parameter of the Weibull distribution which gives approximately the same value to the k parameter (to within 1%) as the Gamma distribution with a shape parameter of 7.

   Compare the metrics $\dfrac{dg}{dt}$ for the two distributions, with the same mean equal to 1000h. Discuss the similarity with Figure 6.2.

5. With the distributions considered in Exercise 4, find an upper bound on the absolute value of the average RUL loss rate, and the range of time for which that upper bound is valid.

6. For the lognormal TTF distribution, does the time transformation $g(t)$ correspond to an S-shaped curve, or is it concave?

7. Derive the reliability function corresponding to a piecewise-linear g-function (for instance with three segments).

BIBLIOGRAPHY

Banjevic, D., "Remaining Useful Life in Theory and Practice", *Metrika*, 69, 337–349, 2009.

Lai, C.D., & Zhang, L., "Mean Residual Life and Other Properties", *International Journal of Reliability, Quality and Safety Engineering*, 11(2), 113–132, 2004.

7 Multiple Failure or Degradation Modes

7.1 INTRODUCTION

So far, the remaining useful life (RUL) or mean residual life (MRL) has been determined under the implicit assumption of one single failure mode. The degradation mechanism, and accordingly the time until failure due to that failure mode, has been modeled by a stochastic process.

In reality, in most applications, several concurrent degradation processes are at work simultaneously, and it is not known in advance which one will lead first to a failure. One is thus confronted with a situation of "competing failure or degradation modes".

After formulating the problem in the general case (Section 7.2) we will then illustrate it in a special case (Section 7.3).

7.2 GENERAL FORMULATION

Consider two failure modes, and the corresponding two times to failure T_1 and T_2 described by the reliability (or survival) functions $R_1(t)$ and $R_2(t)$, respectively.

The time to failure of the system is the minimum of T_1 and T_2.

Then, if the two failure modes are independent, the overall reliability function is equivalent to that of a series system with two components, with respective reliability functions R_1 and R_2:

$$R(t) = R_1(t)R_2(t) \tag{7.1}$$

and the system MRL is obtained as

$$V(t) = \frac{1}{R_1(t) \; R_2(t)} \int_t^\infty R_1(u) \; R_2(u) \, \mathrm{d}u. \tag{7.2}$$

DOI: 10.1201/9781003250685-7

The metric g is expressed (see Chapter 5) as

$$g(t) = \frac{\mu}{k}\{1 - [R_1(t)\ R_2(t)]^{\frac{k}{1-k}}\},$$ (7.3)

where

$$k = \frac{1 - \dfrac{\sigma^2}{\mu^2}}{1 + \dfrac{\sigma^2}{\mu^2}}.$$ (7.4)

And the system expectation μ and variance σ^2 are derived from those of the two subsystems:

$$\mu = P[T_1 < T_2]\,\mu_1 + P[T_2 < T_1]\,\mu_2$$ (7.5)

by conditioning upon the first event (failure of 1 first or failure of 2 first).
Then

$$\sigma^2 = E(T^2) - E(T)^2 = E(T^2) - \mu^2.$$ (7.6)

And,

$$E(T^2) = P[T_1 < T_2]\,.\,E(T_1^2) + P[T_2 < T_1]\,E(T_2^2)$$

$$= P[T_1 < T_2](\sigma_1^2 + \mu_1^2) + P[T_2 < T_1](\sigma_2^2 + \mu_2^2).$$ (7.7)

7.3   ILLUSTRATION IN SPECIAL CASE

Let us consider the case of a component which has a maximum lifetime t_1 and is also subject to random failures, with constant failure rate λ.

This situation can apply, for instance, to some electronic components.

In the absence of the finite-lifetime condition, the distribution of the time to failure would be exponential, with rate λ. But in the situation under study, $T > t_1$ is an event of zero probability:

$$\text{if } t \geq t_1, \text{ then } R(t) = 0.$$

The reliability function of interest is in fact the conditional survival function under the condition that $t \leq t_1$ Indeed:

$$R(t) = P[T > t] = P[T \leq t_1]P[T > t \mid T \leq t_1] + P[T > t_1]\,.\,P[T > t \mid T > t_1]. (7.8)$$

But, since $P[T \leq t_1] = 1$ and $P[T > t_1] = 0$.

$$R(t) = P[T > t \mid T \le t_1] = \frac{P[t < T < t_1]}{P[T < t_1]} = \frac{e^{-\lambda t} - e^{-\lambda t_1}}{1 - e^{-\lambda t_1}} \quad \text{for} \quad t \le t_1 \quad (7.9)$$

and $R(t) = 0$ for $t > t_1$.

Thus the reliability function depends on two parameters: t_1, the maximum life time, and λ, the failure rate. It is plotted in Figure 7.1 for $\lambda = 10^{-3}$ / h and two values of t_1 (Figures 7.1a and 7.1b).

It can be seen that, for t_1 much larger than $1/\lambda$ ($t_1 = 10\ 000$h in the example), the exponential function behavior is dominant. Conversely, for t_1 much smaller than $1/\lambda$ ($t_1 = 100$h in the example), $R(t)$ looks like a uniform distribution over $(0, t_1)$. This can be proved easily as follows.

For convenience, let us denote by A the product λt_1: $A \equiv \lambda t_1$.
Then

$$R(t) = \begin{cases} \dfrac{e^{-\lambda t} - e^{-A}}{1 - e^{-A}} & t \le t_1 \\ 0 & t > t_1 \end{cases}. \qquad (7.10)$$

For $t_1 \gg \dfrac{1}{\lambda}$, i.e. $A \to \infty$, it is seen that $\lim_{A \to \infty} R(t) = e^{-\lambda t}$, i.e. the distribution of T converges to an exponential distribution, as illustrated in Figure 7.1(a).

On the other hand, for $t_1 \ll \dfrac{1}{\lambda}$, i.e. $A \approx 0$, let us see the behavior of $R(t)$:

$$R(t) = \frac{e^{-\lambda t} - e^{-\lambda t_1}}{1 - e^{-\lambda t_1}} = \frac{e^{-A}\left(e^{A - \lambda t} - 1\right)}{1 - e^{-A}} = \frac{e^{A - \lambda t} - 1}{e^A - 1}$$

$$= \frac{1 + A - \lambda t + o(A - \lambda t) - 1}{1 + A + o(A) - 1} = \frac{A - \lambda t + o(A - \lambda t)}{A + o(A)} \approx 1 - \frac{\lambda t}{A}$$

$$= 1 - \frac{t}{t_1} \qquad (7.11)$$

which indeed describes a *uniform* distribution over the interval $(0, t_1)$, as seen in Figure 7.1(b).

(The notation o(x) refers to higher-order terms, i.e. $\lim_{x \to 0} \dfrac{o(x)}{x} = 0$.)

Now the MRL function, $V(t)$, will be derived, as well as the time transformation $g(t)$.

From the definition of MRL,

$$V(t) = \frac{1}{R(t)} \int_t^\infty R(s)\,ds = \frac{1}{R(t)} \int_t^{t_1} R(s)\,ds \quad \text{for } t < t_1 \qquad (7.12)$$

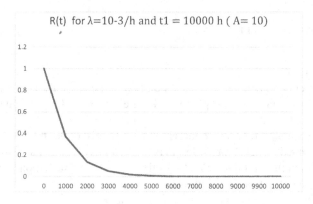

FIGURE 7.1(a)   Reliability function for mixed distribution ($\lambda = 10^{-3}$ / h; $t_1 = 10\,000$h).

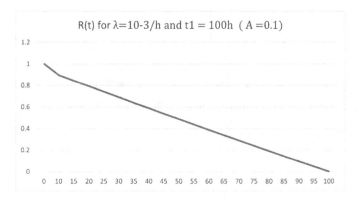

FIGURE 7.1(b)   Reliability function for mixed distribution ($\lambda = 10^{-3}$ / h; $t_1 = 100$h).

$$0 \text{ for } t \geq t_1$$

since $R(s) = 0$ for $s > t_1$.
Thus, from (7.10),

$$V(t) = \frac{1}{R(t)} \int_t^{t_1} \frac{e^{-\lambda s}\, ds}{1 - e^{-\lambda t_1}} - \frac{1}{R(t)} \frac{\left(t_1 - t\right) e^{-\lambda t_1}}{1 - e^{-\lambda t_1}} \quad \text{for } t \leq t_1$$

which leads to

$$V(t) = \frac{1}{\lambda} - \frac{\left(t_1 - t\right)}{e^{\lambda(t_1 - t)} - 1} \quad \text{for } t \leq t_1 \tag{7.13}$$

$$0 \text{ for } \geq t_1.$$

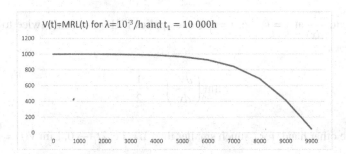

FIGURE 7.2(a) MRL for mixed distribution ($\lambda = 10^{-3} / h$; $t_1 = 10000h$).

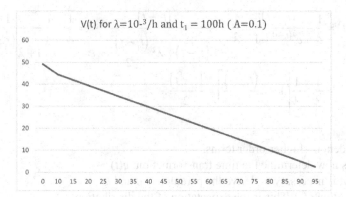

FIGURE 7.2(b) MRL for mixed distribution ($\lambda = 10^{-3} / h$; $t_1 = 100h$).

Notice that $\lim\limits_{t \to t_{1-}} V(t) = 0$ so $V(t)$ is continuous at t_1.

A plot of $V(t)$ is shown in Figure (7.2) for $\lambda = 10^{-3} / h$ and the two values of t_1 previously considered.

For t_1 much larger than $1/\lambda$, i.e. $\lambda t_1 \to \infty$, $\text{Lim } V(t) = \dfrac{1}{\lambda}$, as long as $t \ll t_1$, which indeed corresponds to an exponential distribution for the life time. But for values of t close to t_1, $V(t)$ decreases to 0 and approximates a straight line with slope equal to $-\dfrac{1}{2}$, which is characteristic of the uniform distribution. This can be verified by calculating $\dfrac{dV}{dt}$ from (7.13).

For convenience, set $x = (t_1 - t)$

$$\frac{dV}{dx} = \frac{1 - e^{\lambda x}(1 - \lambda x)}{(e^{\lambda x} - 1)^2}. \tag{7.14}$$

To evaluate $\dfrac{dV}{dx}$ at $x = 0$ (i.e. $t = t_1$), one applies L'Hôpital's rule twice to (7.14), which gives

$$\lim_{x \to 0}\left(\frac{dV}{dx}\right) = -\frac{1}{2}.$$

On the other hand, for t_1 much smaller than $1/\lambda$, i.e. $\lambda t_1 \to 0$, $\mathrm{Lim}\, V(t) = \dfrac{(t_1 - t)}{2}$, which characterizes a uniform distribution over the interval $(0, t_1)$,

Indeed

$$V(t) = \frac{1}{\lambda} - \frac{(t_1 - t)}{e^{\lambda(t_1 - t)} - 1} = \frac{1}{\lambda} - \frac{(t_1 - t)}{\lambda(t_1 - t) + \lambda^2 \dfrac{(t_1 - t)^2}{2} + \varepsilon} \approx \frac{1}{\lambda} - \frac{1}{\lambda}\frac{1}{1 + \lambda\dfrac{(t_1 - t)}{2}}$$

$$\approx \frac{1}{\lambda} - \frac{1}{\lambda}\left[1 - \lambda\frac{(t_1 - t)}{2}\right] = \frac{(t_1 - t)}{2},$$

$$(7.15)$$

where ε denotes higher-order terms.

Let us now determine the time transformation, $g(t)$.

First it is necessary to derive the parameters μ and k.

Derivation of μ. This is the expectation of the distribution:

$$\mu = \mathrm{MTTF} = V(0) = \frac{1}{\lambda} - \frac{t_1}{e^{\lambda t_1} - 1}. \qquad (7.16)$$

From (7.16), it is verified that $\lim\limits_{t_1 \to \infty} \mu = \dfrac{1}{\lambda}$

i.e. μ converges to the expectation of an exponential distribution of rate λ.

Indeed, using the Taylor–McLaurin series expansion of the exponential function,

$$\mu = \frac{1}{\lambda}\left(1 - \frac{\lambda t_1}{(e^{\lambda t_1} - 1)}\right) = \frac{1}{\lambda}\left(1 - \frac{\lambda t_1}{\lambda t_1 + \dfrac{(\lambda t_1)^2}{2} + \dfrac{(\lambda t_1)^3}{3!} + \dots}\right) = \frac{1}{\lambda}\left(1 - \frac{1}{1 + \dfrac{\lambda t_1}{2} + \dfrac{(\lambda t_1)^2}{3!} + \dots}\right).$$

$$(7.17)$$

The same series expansion (7.17) shows that, for $t_1 \ll \dfrac{1}{\lambda}$,

$$\mu \approx \frac{t_1}{2} \tag{7.18}$$

which is the expectation of a uniform distribution over $(0, t_1)$.

Derivation of k

The parameter k will be derived from the variance σ^2 of the time to failure, which we now calculate.

$$\sigma^2 = E\left[(T-\mu)^2\right] = E(T^2) - \mu^2 \tag{7.19}$$

$$E(T^2) = \int_0^\infty t^2(t) f(t) dt \tag{7.20}$$

with

$$f(t) = -\frac{dR}{dt}.$$

In view of (7.10),

$$E(T^2) = \frac{1}{1-e^{-\lambda t_1}} \int_0^{t_1} \lambda t^2 e^{-\lambda t} dt = \frac{1}{\lambda^2(1-e^{-\lambda t_1})} \int_0^{\lambda t_1} u^2 e^{-u} du. \tag{7.21}$$

For convenience, let us define $I(x)$ by

$$I(x) = \int_0^x u^2 e^{-u} du. \tag{7.22}$$

Then

$$E(T^2) = \frac{I(\lambda t_1)}{\lambda^2(1-e^{-\lambda t_1})} = \frac{I(A)}{\lambda^2(1-e^{-A})}. \tag{7.23}$$

Integrating (7.22) by parts leads to:

$$I(x) = 2(1-e^{-x}) - xe^{-x}(x+2). \tag{7.24}$$

There follows:

$$E(T^2) = \frac{2(1-e^{-A}) - Ae^{-A}(A+2)}{\lambda^2(1-e^{-A})} = \frac{2}{\lambda^2} - \frac{Ae^{-A}(A+2)}{\lambda^2(1-e^{-A})}. \tag{7.25}$$

Taking (7.16) into account, one obtains the variance σ^2.

After effecting the products and simplifying, we obtain:

$$\sigma^2 = \frac{1}{\lambda^2} - \frac{A^2 e^{-A}}{\lambda^2 \left(1 - e^{-A}\right)^2}. \tag{7.26}$$

And finally therefore, the square of the coefficient of variation:

$$CV^2 = \frac{\sigma^2}{\mu^2} = \frac{1 - \dfrac{A^2 e^{-A}}{\left(1 - e^{-A}\right)^2}}{\left(1 - \dfrac{Ae^{-A}}{1 - e^{-A}}\right)^2}. \tag{7.27}$$

And the k coefficient is derived in the usual manner (Chapter 5):

$$k = \frac{1 - CV^2}{1 + CV^2}. \tag{7.28}$$

We verify that the limiting cases behave as expected:

For $t_1 \gg \dfrac{1}{\lambda}$ (or A $\to\infty$), Lim CV = 1 and $k \to 0$, consistently with the exponential distribution.

For $t_1 \ll \dfrac{1}{\lambda}$ (or A $\to 0$), Lim $CV^2 = \dfrac{1}{3}$ and Lim $k = \dfrac{1}{2}$, consistently with the uniform distribution.

(This can be verified for instance by applying L'Hôpital's rule.)

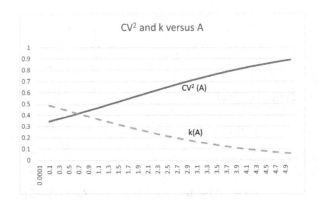

FIGURE 7.3   Square of coefficient of variation, CV^2, and parameter k, as functions of $A = \lambda t_1$.

The plots of CV^2 and k as functions of A are given in Figure 7.3. The value of CV^2 rises from 1/3 to 1, and the value of k decreases from ½ to 0, as A grows from 0 to infinity.

Having determined μ and k, we can now calculate the time transformation $g(.)$ and correspondingly the confidence intervals in application of the general theory of Chapter 5.

$$g(t) = \frac{\mu}{k}\left[1 - R(t)^{\frac{k}{1-k}}\right] = \frac{\mu}{k}\left[1 - \left(\frac{e^{-\lambda t} - e^{-A}}{1 - e^{-A}}\right)^{\frac{k}{1-k}}\right]. \tag{7.30}$$

Also, for the confidence interval, the inverse function is needed:

$$g^{-1}(x) = R^{-1}\left[\left(1 - \frac{k}{\mu}x\right)^{\frac{1-k}{k}}\right]. \tag{7.31}$$

And, from (7.10),

$$R^{-1}(y) = -\frac{\ln\left[e^{-A} + (1 - e^{-A})y\right]}{\lambda}. \tag{7.32}$$

Therefore

$$g^{-1}(x) = -\frac{\ln\left[e^{-A} + (1 - e^{-A})\left(1 - k\frac{x}{\mu}\right)^{\frac{1-k}{k}}\right]}{\lambda}. \tag{7.33}$$

The confidence interval for the RUL can thus be obtained by application of Chapter 5, Theorem 5.2.

The failure rate is obtained directly from (7.10):

$$\lambda(t) = -\frac{\frac{dR}{dt}}{R(t)} = \begin{cases} \dfrac{\lambda e^{-\lambda t}}{e^{-\lambda t} - e^{-A}} & t < t_1 \\ 0, & t \ge t_1 \end{cases}. \tag{7.34}$$

The metric $\dfrac{dg}{dt}$ is illustrated in Figures 7.4(a) and 7.4(b) for two values of t_1.

In Figure 7.4(a), $t_1 = 10000h$ so that A $=10$. This case corresponds to a finite life duration (t_1) much higher than the MTTF of the exponential distribution of parameter λ, which is equal to 1000h. Therefore one can expect the exponential distribution behavior to dominate, at least for small values of the time variable (sufficiently

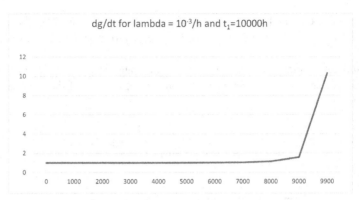

FIGURE 7.4(a) Metric $\dfrac{dg}{dt}$ ($t_1 = 10000$h).

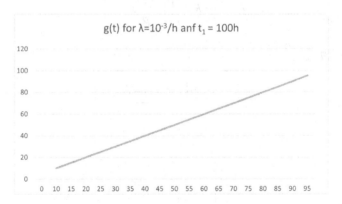

FIGURE 7.4(b) Transformation $g(t)$ ($t_1 = 100$h).

far from the maximum life t_1). This in indeed so: from (7.27) and (7.28), $k = 0.018$, which is close to the value it would have for an exponential time to failure (then it would be strictly equal to 0).

At first, for small values of t, $\dfrac{dg}{dt}$ is constant, close to 1, which corresponds to the exponential distribution case (where $g(t) = t$); but for higher values of t, starting around 9000h, it begins to rise very quicky toward infinity—if there was no failure rate λ but only a finite life time t_1, $\dfrac{dg}{dt}$ would be an impulse function at $t = t_1$.

In Figure 7.4(b), on the contrary, $t_1 = 100$h, so that A $= 0.1$. Then the MTTF of the exponential distribution, 1000h, is much larger than the finite life duration t_1.

Thus the dominant behavior one would expect is that of the Dirac distribution, at least for small values of t. This is visible in the behavior of the failure rate $\lambda(t)$, first close to 0 then sharply rising (Figure 7.5(b)).

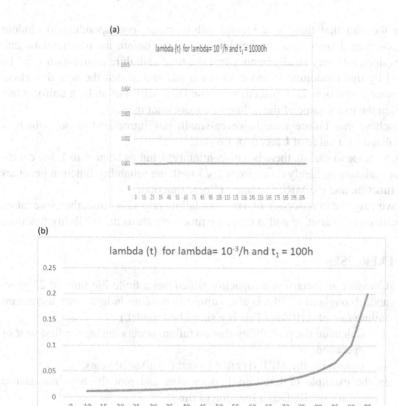

(a)

lambda (t) for lambda= 10⁻³/h and t₁ = 10000h

(b)

lambda (t) for lambda= 10⁻³/h and t₁ = 100h

FIGURE 7.5   Failure rate (a) for $t_1 = 10\,000$h; (b) for $t_1 = 100$h.

The k coefficient is very close to 0.5, which is characteristic of a uniform distribution.

Discussion
In this generic example with two failure modes, it has been shown and proved that, depending on the relative importance of the two failure modes (i.e. the relative values of the MTTF corresponding to random failures, $\dfrac{1}{\lambda}$, and the maximum possible life duration t_1), the reliability function and the MRL are between those corresponding to an exponential distribution (purely random failure with no theoretical lifetime limit) and the uniform distribution with a mean equal to half the finite lifetime.

The latter case corresponds to a maximum lifetime t_1 much smaller than the reciprocal of the random failure rate λ.

It is interesting to note that, if there were no random failures but only a finite lifetime, then the time-to-failure would follow a Dirac distribution (i.e. a deterministic, certain failure at time t_1, and $k = 1$ instead of $k = \dfrac{1}{2}$).

But the fact that there is a second failure mode, corresponding to random failures—even though these are not likely to occur before the deterministic end of life (since λ is very small) implies that the time-to-failure distribution must be affected by that randomness; and calculation has shown that the way this effect takes place is through the replacement of the Dirac distribution by a uniform one (in which the exact value of the failure rate λ does not matter).

Somehow this failure rate, however small, has introduced noise (which is materialized by a value of k less than 1).

If λ. is made to vanish, there is a discontinuity: k jumps from ½ to 1, the coefficient of variation suddenly drops from $1/\sqrt{3}$ to 0, the reliability function becomes a step function and the MRL becomes a linear function.

Conversely, the introduction of a random failure rate adds noise, therefore raises the coefficient of variation and at the same time smoothens the reliability function.

7.4 EXERCISES

1. Consider an electrolytic capacitor which has a finite life time of 25 years and, throughout its life, is also subject to random failures with a constant failure rate of 8.10^{-5}/h. (This is a simplified model.)
   a. Calculate the probability that no failure occurs during the first year of operation.
   b. Determine the MRL (i) after 1 year; (ii) after 20 years.
2. In the example of Exercise 1, determine and plot the 80% confidence interval for the RUL as a function of time.
3. Do Exercises 1 and 2 for a component that has the same failure rate as in Exercise 1 (8.10^{-5}/h) but a much shorter life time, of 2 years (calculate MRL only after 1 year).
4. Redo Exercises 1 to 3 under the assumption of a component with a finite life time of 2 years but no random failures ($\lambda=0$). Compare the results. Why are they so different?

BIBLIOGRAPHY

Ross, S.M., *Introduction to Probability Models*, 10th Ed., Boca Raton, FL, Elsevier, 2010.
Selby, S.M., *Standard Mathematical Tables*, 20th Ed., Cleveland, OH, CRC, The Chemical Rubber Co., 1972.

8 Statistical Estimation Aspects

8.1 INTRODUCTION

The previous chapters focused on probabilistic reasoning. In this chapter, we will be dealing with the question of estimating distributions and estimating remaining useful life (RUL) on the basis of statistical analysis of field data. Field data will typically be run-to-failure data or degradation data.

In Section 8.2, the method of moments is presented. This method estimates the moments of the distributions, see e.g. Van Noortwijk (2009), and Cinlar (1977), without any assumption on the type of probability distribution from which the data originate. This feature makes the method of moments a non-parametric approach (also known as distribution-free methods).

Then, in Section 8.3, a parametric approach, the maximum likelihood estimation (MLE) method, is introduced (Meeker 2021). The MLE method is first applied to the special family of distributions described in Chapter 3, with MRL linear in time. Then, two case studies are analyzed, which require more general distributions. The first case study focuses on the reliability of components (LEDs), while the second application demonstrates the analysis method on a redundant system.

An alternative method to determine the g transformation directly from data is presented in Section 8.4. Finally, Section 8.5 gives a brief overview of the Bayesian estimation method, with application to the distributions with MRL linear in time.

8.2 NONPARAMETRIC ESTIMATION

First, let us imagine a simple "toy" example, where a number of run-to-failure data have been collected. It is desired to estimate the reliability function $R(t)$, the mean residual life $V(t)$, and the time transformation $g(t)$. The data are summarized in Table 8.1.

A sample of $n = 9$ failures was observed (for n distinct units), and the corresponding failure times are denoted $t_1, t_2, \ldots, t_n$.

The time axis is divided into equal intervals of length 25h from $t = 0$ to $t = 550$h, and the RUL at each time point is estimated for unit i, as:

DOI: 10.1201/9781003250685-8

TABLE 8.1
Experimental Failure Data

Unit #i	Time of Failure (h) t_i
1	190
2	100
3	150
4	300
5	210
6	50
7	250
8	500
9	150

$$RUL_i(t) = \begin{cases} t_i - t, & t_i > t \\ 0, & t_i \le t \end{cases}$$

or equivalently

$$RUL_i(t) = \max(t_i - t; 0). \tag{8.1}$$

From now on, lower case letter will be utilized to denote statistical estimates (also called empirical values), while the random variables to be estimated are denoted by capital letters.

Thus the empirical MRL, denoted $v(t)$, is obtained as:

$$v(t) = \frac{1}{n} \sum_{i=1}^{i=n} \max(t_i - t; 0). \tag{8.2}$$

The empirical reliability function, $r(t)$, is obtained as the proportion of items still operational at time t:

$$r(t) = \frac{1}{n} \sum_{i=1}^{i=n} I_i(t) \tag{8.3}$$

where by definition

$$I_i(t) = \begin{cases} 1, & t_i > t \\ 0, & t_i \le t \end{cases}. \tag{8.4}$$

The empirical mean time to failure, m, is obtained as

$$m = \frac{1}{n}\sum_{i=1}^{i=n} t_i. \tag{8.5}$$

The coefficient k (cf. Chapters 5 to 7) can be estimated from the coefficient of variation cv:

$$\hat{k} = \frac{1-cv^2}{1+cv^2} \tag{8.6}$$

and the square of the coefficient of variation CV^2 can be estimated (as cv^2) from the estimated mean m and the estimate s^2 of the variance:

$$s^2 = \frac{1}{(n-1)}\sum_{i=1}^{i=n}(t_i - m)^2. \tag{8.7}$$

However, although m is an unbiased estimate of the mean and s^2 is an unbiassed estimate of the variance, the ratio $\frac{s}{m}$ is not an unbiased estimate of the coefficient of variation (Ye 2022).

Be that as it may, the computations are performed as indicated and lead to an estimation $\hat{g}(t)$ of the time transformation $g(t)$ by

$$\hat{g}(t) = \frac{m}{\hat{k}}\left(1 - r(t)^{\frac{\hat{k}}{1-\hat{k}}}\right). \tag{8.8}$$

From this estimate of $g(t)$, is obtained an estimate $v'(t')$ of the MRL $V'(t')$ in the transformed space, as follows:

$$v'(t') = v'\left(\hat{g}(t)\right) = m - \hat{k}\hat{g}(t) \tag{8.9}$$

The results are summarized in Tables 8.1(a) and 8.1(b).

The estimated function $\hat{g}(t)$ is plotted in Figure 8.1; the estimated MRL, $v(t)$, in Figure 8.2; and finally the estimated MRL in the transformed space, $v'(t')$, in Figure 8.3.

Note that the function $v'(t')$ should be close to linear, by definition.

The function v' can also be derived directly from the data instead of through (8.9), i.e., through (8.2) where t_i is replaced with $\hat{g}(t_i)$, $i = 1, \ldots n$.

The two plots are both shown in Figure 8.3. The plot built directly from the data (dotted line) is less close to linearity than the one derived from the equation (solid line): some extrapolations had to be made when no data was available for given time values.

TABLE 8.1(a)
Empirical Estimations

t		0	100	200	300	400	500
m	211.11						
s^2	17386						
$\hat{k}$	0.438741						
$v(t)$		211.11	116.67	51.11	22.22	11.11	0
$r(t)$		1	0.78	0.44	0.11	0.11	0
$\hat{g}$		0	85.82	225.91	394.80	394.80	481.17

TABLE 8.1(b)
Estimation of MRL in Transformed Space: $v'(t')$

t'	0	42	85	177	226	277	333	395	481
$v'(t')$ from $g(t)$	211	193	173	133	112	89	65	38	0
$v'(t')$ direct estimation	238	196	157	86	61	42	23	10	0

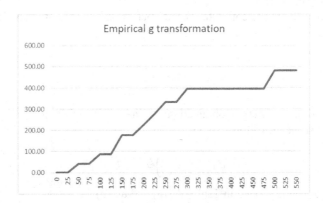

FIGURE 8.1 Estimate of $g(.)$ transformation in the example.

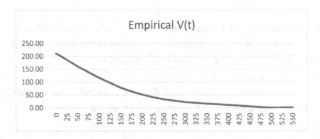

FIGURE 8.2 Estimate of MRL function in the example.

8.3 PARAMETRIC ESTIMATION—ILLUSTRATION ON TWO CASE STUDIES

In this section, a parametric estimation method, the MLE, is reviewed, and then demonstrated in two case studies. First, the MLE principle and methods are recalled and applied to the family of distributions described in Chapter 3. Then, two use cases are studied.

8.3.1 PARAMETRIC ESTIMATION: THE MAXIMUM LIKELIHOOD ESTIMATION METHOD

There are several parametric estimation methods, i.e. methods for estimating an unknown parameter (such as the mean, or the variance) in a parametric model (for instance, a Weibull or a Gamma distribution).

One of the most successful methods—see e.g. (Meeker 2021), is the MLE. Other methods include the least-squares estimation, i.e. linear regression.

We now succinctly describe the MLE method and explain its desirable features.

8.3.1.1 Maximum Likelihood Estimator

Let us denote by θ the unknown parameter which is to be estimated. The notation θ can refer to a scalar, if there is a single unknown parameter, or a vector if there is more than one parameter to be estimated. For instance, if we need to estimate both the mean, and the variance: then the 2D vector θ stands for the two parameters (θ_1, θ_2); or, more generally, for m parameters to be estimated, the m-dimensional vector $\underline{\theta}$:

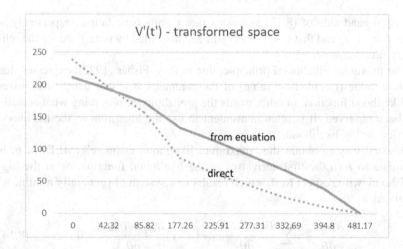

FIGURE 8.3 Estimate of MRL function in transformed space (two methods).

$$\underline{\theta} = \left(\theta_1, \theta_2, \ldots \theta_m\right). \tag{8.10}$$

Suppose n units are observed during a time T, and the failure times of the n units are denoted $t_1, t_2, \ldots t_n$, respectively. Let us first consider the case of a complete *sample*, i.e. one where all units fail before the end of observations,:

$$t_i < T \; \forall \; i.$$

The *likelihood function* is then defined as:

$$L\left(t_1, t_2, \ldots, t_n \middle| \boldsymbol{\theta}\right) \equiv f\left(t_1, t_2, \ldots t_n \middle| \boldsymbol{\theta}\right) \tag{8.11}$$

i.e. the probability density function, evaluated at the times of failure that have been observed.

Usually, the n units are stochastically independent. In that case, the rules of probability calculus for independent events imply that:

$$f\left(t_1, t_2, \ldots t_n \middle| \boldsymbol{\theta}\right) = f\left(t_1 \middle| \boldsymbol{\theta}\right) f\left(t_2 \middle| \boldsymbol{\theta}\right) \ldots f\left(t_n \middle| \boldsymbol{\theta}\right). \tag{8.12}$$

If some of the units have not failed during the observation period, i.e. the observations are right-censored, then, denoting 1, 2,... r the r units ($r \leq n$) that have failed up to time T, out of the n units of the sample, the likelihood function is expressed as:

$$L\left(t_1, t_2, \ldots t_n, \middle| \boldsymbol{\theta}\right) = f\left(t_1 \middle| \boldsymbol{\theta}\right) f\left(t_2 \middle| \boldsymbol{\theta}\right) \ldots f\left(t_r \middle| \boldsymbol{\theta}\right) R\left(T \middle| \boldsymbol{\theta}\right)^{n-r}. \tag{8.13}$$

The right-hand side of (8.13) expresses that r units have failed, respectively at times $t_1, t_2, \ldots t_r$, and that $(n - r)$ units still had not failed by time T (hence the reliability function appears).

The maximum likelihood principle, due to R.A. Fisher (1922), expresses that the best value (i.e. likeliest value) of the parameter $\boldsymbol{\theta}$ is that which maximizes the likelihood function, in other words the probability of observing what actually has been observed. It is often convenient to use the logarithm of the likelihood function, or log-likelihood.

A usual way to obtain the "maximum likelihood estimate" (MLE) of $\boldsymbol{\theta}$, is to equate to zero the first derivative of the likelihood function, or of the log-likelihood, with respect to $\boldsymbol{\theta}$, which results in a system of (generally nonlinear) equations in θ:

$$\frac{\partial \log L}{\partial \theta_j} = \sum_{i=1}^{r} \frac{\partial \log f(t_i \middle| \boldsymbol{\theta})}{\partial \theta_j} + (n-r) \frac{\partial \log R(T \middle| \boldsymbol{\theta})}{\partial \theta_j} = 0 \tag{8.14}$$

$j = 1, \ldots m$

which is a system of m equations in the m unknown parameters $\theta_1, \theta_2 \ldots \theta_m$. If there are no particular constraints on the parameters θ, the maximum likelihood estimator (MLE), $\hat{\theta}$, is a solution of that system. It is a function of the observations:

$$\hat{\theta} = \hat{\theta}\left(t_1, t_{2,} \ldots t_n, T\right). \tag{8.15}$$

(An example is shown below of a parameter which is constrained to belong to a certain set and where the global maximum is not obtained by just solving (8.14).)

For instance, it is easy to show that, if the data follow an exponential distribution (with therefore, one parameter to be estimated, the mean μ), the MLE $\hat{m}$ for μ is

$$\hat{m}\left(t_1, t_2, \ldots t_n, T\right) = \frac{\sum_{i=1}^{r} t_i}{r} + \frac{n-r}{r}T. \tag{8.16}$$

The estimator $\hat{m}$, or the estimator $\hat{\theta}$ in general, is a function of the random data which are collected; it is therefore a random variable, and usually has a mathematical expectation. In the above example,

$$E\left(\hat{m}\right) = \mu + \frac{n-r}{r}T. \tag{8.17}$$

Therefore, if the sample is complete ($r = n$), the expectation of the sample mean $\hat{m}$ is equal to the distribution expectation μ: the estimator $\hat{m}$ is then called *unbiased*.

In general, the *bias*, i.e. the difference between the expectation of the estimator and the parameter to be estimated, is equal to the term $\frac{n-r}{r}T$, which shows that it grows with the amount of censoring (the proportion of items that did not fail before T). In general indeed, censoring introduces a bias.

On the other hand, the variance of the estimator decreases as the sample size increases.

8.3.1.2 Desirable Properties of the Maximum Likelihood Estimator

The MLE has the following properties, which make it a desirable estimator: (see e.g. Meeker 2021 or Pham 2021)

1. It is consistent: it converges to the true value as the sample size goes to infinity.
2. It is asymptotically unbiased.
3. It is asymptotically efficient, i.e. its variance is smallest among all estimators.
4. It is asymptotically normally distributed.

8.3.1.3   Derivation of a Confidence Interval

A confidence interval for the MLE estimator can be obtained from the Fisher information measure. The derivative $\dfrac{\partial \log f}{\partial \theta}$ is called the score. (see e.g. Meeker 2021 or Pham 2021)

The Fisher information measure is defined as the variance of the score. In fact:

$$I(\theta) = E\left[-\frac{\partial^2 \log f}{\partial \theta^2} \right]. \tag{8.18}$$

This expression results from the fact that the expectation of the score is 0, from (8.14).

It is used in order to build a confidence interval around the MLE, by means of a limited Taylor expansion.

In the multi-parameter case, the Fisher information matrix is defined, analogously, as the covariance matrix of the score, and the following holds :

$$I_{ij}(\boldsymbol{\theta}) = E\left[-\frac{\partial^2 \log f}{\partial \theta_i \, \partial \theta_j} \right]. \tag{8.19}$$

A number of commercial statistical packages exist, which implement those methods. One of those has been used in the next two sections (Reliasoft Weibull++, of HBM Prenscia).

The method is now illustrated on the special family of TTF probability distributions with MRL linear in time introduced in Chapter 3; subsequently, in Section 8.3.2, the method will be applied to more general probability distributions, when discussing the use cases.

For the special family introduced in Chapter 3, the probability density is given by (3.18):

$$f(t;k;\mu) = \frac{(1-k)\mu^{1-\frac{1}{k}}}{(\mu - kt)^{2-\frac{1}{k}}} = \frac{1-k}{\mu}\left(1 - \frac{kt}{\mu}\right)^{\frac{1}{k}-2} \quad \text{for } 0 \leq t \leq \frac{\mu}{k}$$

$$f(t;k;\mu) = 0 \qquad\qquad\qquad\qquad \text{for } t \leq 0 \text{ or } t > \frac{\mu}{k} \tag{8.20}$$

and the reliability function, by (3.7):

$$R(t;k,\mu) = \begin{cases} \left(\dfrac{\mu}{\mu - kt}\right)^{1-\frac{1}{k}} = \left(1 - \dfrac{kt}{\mu}\right)^{\frac{1}{k}-1} & 0 \leq t \leq \dfrac{\mu}{k}. \\ 0 & t > \dfrac{\mu}{k} \end{cases} \tag{8.21}$$

Thus, the likelihood function, for r failures at times $t_1, t_2, \ldots, t_r$ and $(n-r)$ survivals after a period T for a sample of size n, can be expressed as follows:

$$L(t_1, t_2, \ldots, t_r, \ldots, T; \mu, k) = \left(\frac{1-k}{\mu}\right)^r \prod_{i=1}^{i=r} \left(1 - \frac{kt_i}{\mu}\right)^{\frac{1}{k}-2} \left(1 - \frac{kT}{\mu}\right)^{(n-r)\left(\frac{1}{k}-1\right)}. \tag{8.22}$$

In case of a complete sample ($r = n$),

$$L(t_1, t_2, \ldots t_n; \mu, k) = \left(\frac{1-k}{\mu}\right)^n \prod_{i=1}^{i=n} \left(1 - \frac{kt_i}{\mu}\right)^{\frac{1}{k}-2}. \tag{8.23}$$

To facilitate calculations, let us denote by θ the ratio k/μ:

$$\theta \equiv \frac{k}{\mu}. \tag{8.24}$$

Thus, over the support of the random variable "time to failure",

$$0 \leq \theta \leq 1/\mu.$$

With this notation, the likelihood function is expressed as a function of θ and k:

$$L(t_1, t_2, \ldots t_r, \ldots, T; \theta, k) = \theta^r \left(\frac{1}{k} - 1\right)^r \prod_{i=1}^{i=r} (1 - \theta t_i)^{\frac{1}{k}-2} \left(\frac{1}{(1-\theta T)}\right)^{(n-r)\left(1-\frac{1}{k}\right)} \tag{8.25}$$

Let us focus on the complete sample case ($r = n$):

$$L(t_1, t_2, \ldots t_n; \theta, k) = \theta^n \left(\frac{1}{k} - 1\right)^n \prod_{i=1}^{i=n} (1 - \theta t_i)^{\frac{1}{k}-2}. \tag{8.26}$$

Therefore the log-likelihood is then:

$$\mathrm{Log}L(t_1, t_2, \ldots t_n; \theta, k) = n \log \theta + n \log\left(\frac{1}{k} - 1\right) + \left(\frac{1}{k} - 2\right) \sum_{i=1}^{i=n} \log(1 - \theta t_i). \tag{8.27}$$

A simple example is provided by the failure time data in Table 8.2. Those data are typical of a uniform distribution, therefore we expect to find $k = \frac{1}{2}$.

The corresponding log-likelihood function is displayed in Table 8.3, as a function of k and $1/\theta$ (in h) and illustrated in Figure 8.4 as a 3D plot.

The table contains only values of θ such that $\frac{1}{\theta} \geq 550$h, with a margin above the observed maximum TTF value, which is 500h.

TABLE 8.2

Example of Failure Times (Sample Size: 5)

Item Number	t_i (h)
1	100
2	200
3	300
4	400
5	500

It is seen that the likelihood function L reaches its maximum for $k = \frac{1}{2}$, as expected, and $1/\theta = 550h$. The maximum value of the log-likelihood is equal to -31.55.

Therefore the parameter estimates are:

$$\hat{k} = 1/2;$$

$$\hat{\theta} = 1/550h^{-1}$$

and the estimate of the mean is $\hat{\mu} = \dfrac{\hat{k}}{\hat{\theta}} = 225h$. Plots of the likelihood function L as a function of k, for several values of θ, are displayed in Figure 8.5.

Note that, since a random variable with MRL linear in time has finite support ($t \leq \mu/k$), the following inequality holds: $\dfrac{1}{\theta} \geq t_i$ for all $i = 1,\ldots n.$, which is why the table is limited to values of $\dfrac{1}{\theta}$ greater than, or equal to $\max(t_i, i = 1,\ldots n) = 500h$.

Therefore the maximum value of the likelihood (or log-likelihood) function does not necessarily correspond to a vanishing value of the derivatives, as the maximum must necessarily belong to the domain defined by $\dfrac{1}{\theta} \geq \max(t_i)$, $i = 1,\ldots n.$, and it can be reached for values of θ on the boundary of the domain instead of in its interior.

In fact, for $k = 1/2$, (8.27) shows that $\log L = n \log\theta = -n \log\left(\dfrac{1}{\theta}\right)$ whose maximum corresponds to the smallest possible value of $\log\left(\dfrac{1}{\theta}\right)$, i.e. also the smallest allowable value of $\dfrac{1}{\theta}$.

Note however that, for values of $1/\theta$ close to $500h$, i.e. the maximum recorded TTF value among the five samples, the log-likelihood function diverges to $-\infty$, or the likelihood function converges to 0, since $(1 - \theta t_i)$ vanishes for one value of the

TABLE 8.3
Log-Likelihood Function for the Example of Table 8.2

1/θ	0.1	0.2	0.3	0.4	0.5	0.55	0.6	0.7	0.8	0.9
550	-61.67	-40.03	-34.16	-32.09	-31.55	-31.62	-31.86	-32.85	-34.63	-37.97
600	-54.37	-37.57	-33.31	-32.04	-31.98	-32.23	-32.62	-33.84	-35.79	-39.26
650	-50.00	-36.18	-32.92	-32.15	-32.38	-32.74	-33.22	-34.58	-36.63	-40.19
700	-46.97	-35.27	-32.72	-32.30	-32.76	-33.19	-33.73	-35.19	-37.32	-40.94
750	-45.95	-35.11	-32.84	-32.56	-33.10	-33.56	-34.13	-35.63	-37.80	-41.44
800	-42.96	-34.19	-32.61	-32.68	-33.42	-33.96	-34.60	-36.19	-38.43	-42.13
850	-41.56	-33.85	-32.63	-32.87	-33.73	-34.30	-34.97	-36.62	-38.89	-42.62
900	-40.41	-33.60	-32.67	-33.07	-34.01	-34.62	-35.31	-37.01	-39.31	-43.07
950	-39.46	-33.41	-32.74	-33.27	-34.28	-34.92	-35.64	-37.36	-39.70	-43.47
1000	-38.67	-33.27	-32.82	-33.46	-34.54	-35.20	-35.94	-37.70	-40.05	-43.85
1050	-37.99	-33.17	-32.91	-33.64	-34.78	-35.46	-36.22	-38.01	-40.38	-44.19
1100	-37.41	-33.10	-33.01	-33.82	-35.02	-35.71	-36.49	-38.30	-40.69	-44.51
1150	-36.91	-33.05	-33.11	-34.00	-35.24	-35.95	-36.74	-38.57	-40.98	-44.82
1200	-36.48	-33.02	-33.22	-34.17	-35.45	-36.18	-36.98	-38.83	-41.26	-45.10
1250	-36.10	-33.01	-33.32	-34.34	-35.65	-36.40	-37.21	-39.07	-41.51	-45.37
1300	-35.77	-33.01	-33.43	-34.50	-35.85	-36.61	-37.42	-39.31	-41.76	-45.63
1350	-35.47	-33.02	-33.54	-34.66	-36.04	-36.81	-37.63	-39.53	-41.99	-45.87
1400	-35.22	-33.03	-33.65	-34.82	-36.22	-37.00	-37.83	-39.74	-42.22	-46.10
1450	-34.99	-33.06	-33.76	-34.97	-36.40	-37.18	-38.02	-39.95	-42.43	-46.32
1500	-34.79	-33.09	-33.86	-35.11	-36.57	-37.36	-38.21	-40.14	-42.63	-46.53

k

Log Likelihood

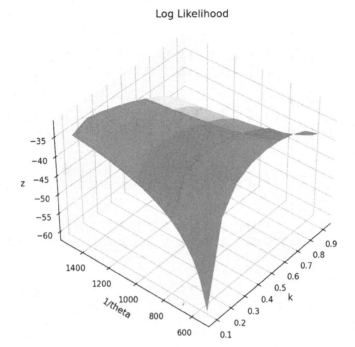

Log Likelihood

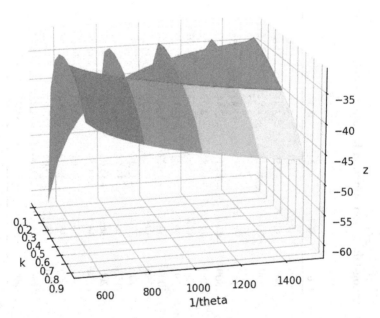

FIGURE 8.4   Log-likelihood as a function of k and $\frac{1}{\theta}$.

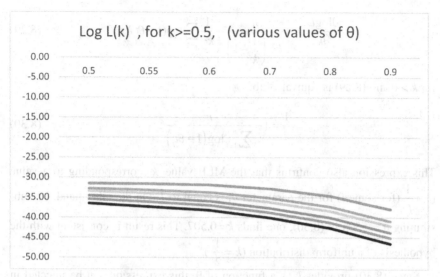

FIGURE 8.5 Log-likelihood as a function of k and θ (from top: $\dfrac{1}{\theta}$ = 500h; to bottom: $\dfrac{1}{\theta}$ = 1500h); ($k \geq \dfrac{1}{2}$).

index i. For such values of θ, k converges to 1. Remembering the meaning of $k = 1$, this would correspond to a deterministic (Dirac) distribution instead of a uniform one ($k = 1/2$).

Careful observation of Table 8.3 also shows that the log-likelihood function has a local maximum corresponding to $k = 0.2$ and $\dfrac{1}{\theta} = 1300h$, but that local maximum (equal to −33.01) is smaller than the global maximum (equal to −31.55).

Thus an algorithm that sets the derivatives to zero as in (8.14) without taking any other constraint into account might converge to that local maximum and lead to an erroneous estimate. In dealing with the derivation of MLE estimate, it is desirable to use all the resources of optimization theory, and in particular the Kuhn–Tucker conditions (see e.g. Luenberger 1968), which take the inequality constraints into account and introduce a Lagrange multiplier. Also of course, it is wise, when selecting the solution, to draw on the physical knowledge which the engineer has on the physics of the system,

Equation (8.14) in the special case of a distribution with linear MRL takes the following form:

$$\frac{\partial \log L}{\partial \theta} = \frac{n}{\theta} - \left(\frac{1}{k} - 2 \right) \sum_{i=1}^{i=n} \frac{t_i}{1 - \theta t_i} = 0 \qquad (8.28)$$

$$\frac{\partial \log L}{\partial k} = -\frac{n}{k^2 \left(\dfrac{1}{k}-1\right)} - \frac{1}{k^2}\sum_{i=1}^{i=n}\log(1-\theta t_i) = 0 \tag{8.29}$$

for $k > 0$ and (8.29) is equivalent to:

$$\frac{1}{k} = 1 - \frac{n}{\sum_{i=1}^{n}\log\left(1-\theta t_i\right)}. \tag{8.30}$$

This expression also confirms that the MLE value, $\hat{k}$, corresponding to a value $\hat{\theta} = \dfrac{1}{t_i}$ (for some i) for the parameter θ, is equal to 1. On the other hand, by substituting $\theta = 550$h in (8.30), one finds $k = 0.507$. This result is consistent with the hypothesis of a uniform distribution ($k = \dfrac{1}{2}$).

Since (8.30) provides k as a function of θ, this expression can be injected in (8.28), resulting in a nonlinear equation in θ only, which can be solved by standard numerical methods. Upon doing so, it is verified for instance that, in the degenerate case when $t_i = t*$ for all i (i.e. all devices fail at the same time t_i), the solution to (8.28) and (8.29) is $1/\theta = t*$ and $k = 1$, which is the "Dirac" distribution corresponding to a deterministic lifetime of $t*$.

Equation (8.28) shows that the behavior of the likelihood function as a function of θ will be markedly different depending on the position of k with respect to ½. This is also apparent from (8.27).

For $k \geq 1/2$, L is a monotonically increasing function of θ (Figures 8.4 and 8.5). Thus, for $k \geq 1/2$, the maximum value of the log-likelihood function is found for the smallest value of $1/\theta$, i.e. 550h, and it is a global maximum. Such is not the case for $k < \dfrac{1}{2}$ and therefore, for values of k in that range, there can be local maxima, as seen above for $k = 0.2$.

In conclusion of this example, the calculation of the MLE based on the sample of Table 8.2 would yield $\dfrac{1}{\hat{\theta}} = 550$h, $\hat{k} = \dfrac{1}{2}$, therefore $\hat{\mu} = 225$h.

Finally, note that, if the method of moments of Section 8.2 is used in this example, the following estimates are obtained: $\hat{\mu}=300$h; $\hat{k} = 0.56$. The estimation is less precise for k but provides the arithmetic mean for the estimate of μ.

8.3.2   First Case Study: Light-Emitting Diodes (LEDs)

Light-emitting diodes (LEDs) are increasingly used in industry because of their energy efficiency and reliability, and not only in consumer and public applications, such as indoor and outdoor lighting, but also in industrial applications. An example of industrial application is for optocouplers where, for safety reasons and to avoid voltage-induced damage to sensitive electronic equipment, the emitting current is first converted into light by an LED and later converted back into a current using the photoelectric effect.

Due to various stresses, the efficiency of LEDs, i.e. the amount of light generated per unit of input current, degrades over time (due to a deterioration of the quantum efficiency). The prominent stress factors affecting the efficiency of LEDs are input current intensity and temperature, and physics-based degradation models, such as Black's law (Black 1967), have been devised to model the influence of those stressors on the performance of LEDs. Note that the effect of accelerating factors is, often, highly nonlinear. A recent European research project, known as the AI-powered Digital twin for lighting infrastructure in the context of front-end Industry 4.0, "AI-TWILIGHT", has studied the reliability of LEDs. Accelerated life tests have been extensively applied in the AI-TWILIGHT project, and several LED types were analyzed under various combinations of current and temperature, well above the normal operating conditions. The effects of the stresses on the LEDs are multiple and include luminous flux reduction and color shift. Here we shall focus on the flux reduction effect (Rocchetta 2022).

In this work, the failure of an LED is defined as the crossing of a threshold equal to 98% of the nominal luminous flux. Note that this is a relatively harsh requirement compared to the industry best practice, which defines failure of LED at 70% (for indoor applications) or 90% (for a few outdoor applications) depreciation levels. With this definition of failure, times to failure data are gathered for various test samples. The accelerated degradation experiment applies two levels of accelerated forward current intensities, set to 0.35 A and 0.7 A, and temperatures, 85°C and 105°C. This experimental setup leads to four combinations of stress levels, each defined by a level of current and temperature. The observed time-to-failure data are reported in Table 8.4. Note that some interpolation was needed as the recorded raw data did not always show the exact time of threshold crossing. The sample size is 25 for each stress combination (25 units under test). As expected, for some stress combinations, no failure had been observed by the end of the test, i.e. the failure time was right-censored. Table 8.4 shows the right-censored data in a bold font, i.e., the end-of-test and not a failure time. For instance, for the stress combination $T = 105°C$ and $I = 0.35A$, the value of 9072h corresponds to the end of observations. In other words, unit 15 survived until time $t = 9072h$ when the test terminated. Also note that for the stress combination $T = 85°C$ and $I = 0.35A$ a subset of LED units are tested for 6048h, while for other combinations, they were tested up to 9072h, which explains the difference in right-censoring times.

Several probability distributions were fitted to the data and the best fit was selected from among a large number of distribution families: Weibull (two and three parameters), exponential (one and two parameters), gamma, lognormal, Gumbel, and so on. The criterion used for model selection was the minimum Akaike information criterion (AIC). The AIC index is defined as follows:

TABLE 8.4
Results of LED Accelerated Tests for 25 Samples

Temperature [°Celsius]	105	85	85	105
Current [A]	0.35	0.7	0.35	0.7
Sample number				
1	8960	6249.6	5040	5156
2	6552	4872	6048	4032
3	2419.2	4435.2	3931.2	2520
4	3738	4872	**9072**	3969
5	2772	4284	5040	2898
6	3024	4620	4284	3456
7	1512	2072	6300	2419.2
8	2520	5544	4536	4032
9	8883	7056	5670	4032
10	2688	4788	3528	4939.2
11	4032	5846.4	**9072**	2337
12	2772	4284	6048	2352
13	1915.2	4636.8	6048	2016
14	2985.2	4872	5796	2337
15	**9072**	4788	**9072**	2310
16	2978.2	4704	3024	2520
17	6384	2973.6	**9072**	2841
18	3024	4737.6	6006	2822.4
19	2940	4599	4444	2474
20	8412.9	4536	**6048**	2184
21	5472	5773	**6048**	2142
22	5334	3940	**6048**	3528
23	3402	1512	**6048**	2952
24	3024	3960	**6048**	2795
25	5040	1512	**6048**	2464

$$\text{AIC} = -2\log(\max L) + 2P. \qquad (8.31)$$

In (8.31), the term $\log(\max L)$ denotes the maximum value of the log-likelihood function, and P the number of parameters of the model (Kotz 1992). The reason for including P is to penalize models that involve too many parameters. For instance, for the three-parameter Weibull distribution, $P = 3$, and max L is the likelihood function evaluated at $\eta = \eta^*$, $\beta = \beta^*$, $\gamma = \gamma^*$, where $(\eta^*, \beta^*, \gamma^*)$ is the MLE.

Note once again that physical knowledge is important in selecting a model. For instance, in the simple example of Table 8.3, the AIC criterion would have led to selecting a two-parameter exponential distribution (shifted exponential) since the corresponding maximum log-likelihood is equal to −31.49, marginally better than the one found in 8.3.1 (−31.55) and therefore also a higher AIC since the number

TABLE 8.5
Results of Statistical Parameter Estimation for LEDs

Applied Stresses		μ/k	MTTF = μ (h)	k	Inflection Point (h)	Distribution	η or Scale(h)	β or Shape
I (A)	T (°C)	(h)						
0.35	85	9113	6826	0.749		Gamma	6.89	6.98
		9009	6757	0.75	4455	2P Weibull	7581	2.86
0.35	105	7556	4458	0.59		Gamma	7.04	3.88
0.7	85	5313	4463	0.84	2940	2P Weibull	4934	3.85
0.7	105	3678	3016	0.82	1982	2P Weibull	3347	3.58

TABLE 8.6
Estimated MTTF as a Function of T and I

MTTF (h)		I (A)	
		0.35	**0.7**
T (°C)	**85**	6826	4463
	105	4458	3016

Note: Values of **T** and **I** are in bold figures.

TABLE 8.7
Estimated k as a Function of T and I

k		I (A)	
		0.35	**0.7**
T (°C)	**85**	0.749	0.84
	105	0.59	0.82

Note: Values of **T** and **I** are in bold figures.

of parameters is 2 in both cases. But the exponential distribution implies a constant MRL, rather than one which decreases with a slope of ½.

In general, it is necessary to perform a goodness-of-fit test—see, e.g., (Birolini 2017; Di Bucchianico 2022), in order to test the hypothesis that the data fit an assumed distribution.

The results of the analysis are summarized in Table 8.5.

The best-fit distribution according to the AIC is either the two-parameter Weibull or the Gamma distribution, depending on the test case. Even where the Gamma distribution maximizes the AIC, the 2P Weibull distribution is not very far, which is why we have also included it. The inflection point of the g transformation was calculated with that distribution. Tables 8.6 and 8.7 summarize the dependence of the MTTF and the k parameter on the two applied stresses, T and I; the same is illustrated in Figures 8.6 and 8.7.

For the estimation of k, the formulas given in Chapter 5 have been used.

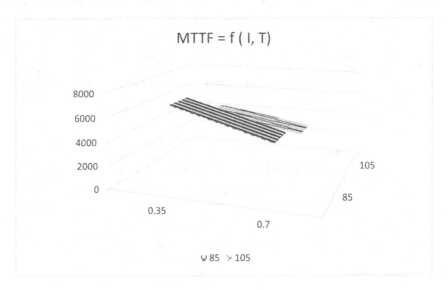

FIGURE 8.6   Estimated MTTF as a function of stresses T and I.

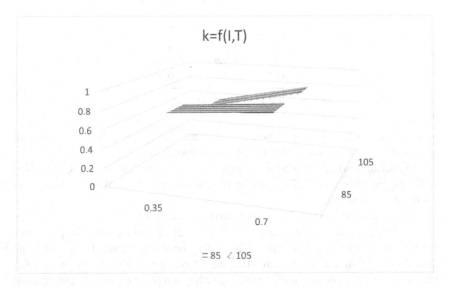

FIGURE 8.7   Estimated k parameter as a function of stresses T and I.

For the two-parameter Weibull distribution,

$$k = \frac{1-\left(\dfrac{\sigma}{\mu}\right)^2}{1+\left(\dfrac{\sigma}{\mu}\right)^2} = \frac{2\Gamma^2\left(1+\dfrac{1}{\beta}\right)}{\Gamma\left(1+\dfrac{2}{\beta}\right)} - 1 \tag{8.32}$$

and, for the Gamma distribution,

$$k = \frac{\beta-1}{\beta+1}. \tag{8.33}$$

Note that β designates different values in (8.32) and (8.33), i.e. the shape factor of the Weibull and Gamma distribution, respectively. Table 8.3 shows that the Gamma and Weibull fit lead to about same estimation for k.

Likewise, the MTTF is calculated from the scale and shape parameters as described in Chapter 5.

As the maximum likelihood method allows for the derivation of confidence intervals (using for instance the Fisher information matrix), confidence bounds for k and MTTF can be obtained correspondingly.

As expected, higher stresses, in temperature or current, reduce the TTF: the luminous flux reaches the 98% threshold earlier. Thus the MTTF drops from a value of 6826h for $I = 0.35$ A and $T = 85°$C, to a value of only 3016h for I = 0.7 A and T = 105 °C.

Also, the parameter k increases with higher current but not necessarily with higher temperature. It is visible from Tables 8.6 and 8.7 and Figures 8.6 and 8.7 that the relative impact of the two stressors is quite variable—for instance, the effect of temperature on k is almost negligible at high current. We observe that while a temperature rise lowers the MTTF, a rise in current lowers the coefficient of variation. In other words, higher current levels reduce the uncertainty affecting the failure time.

Let us study the reliability function, the g function, and the MRL for some of those examples.

For the lowest stress level ($I = 0.35$ A, $T = 85$ °C), the reliability function $R(t)$ is plotted in Figure 8.8, and the MRL in Figure 8.9, with a 90% confidence interval. The time transformation $g(t)$ and its derivative dg/dt are plotted in Figures 8.10 and 8.11, respectively.

Using the formula of Chapter 6 (6.26), the inflection point of $g(t)$ is found to be at $t^*=4455$h, as seen on the plot. According to Theorem 6.1 (Chapter 6), this means that, for $t > 4455$h,

$$\left|\frac{d\mathrm{MRL}}{dt}\right| \leq 0.749.$$

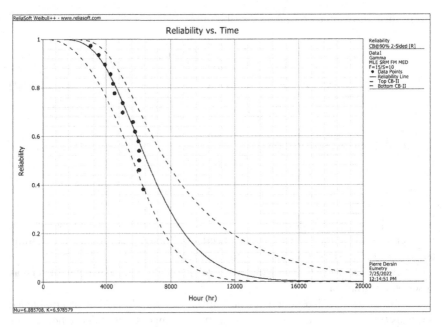

FIGURE 8.8   Reliability function for LED (I = 0.35 A, T = 85°C) with 90% confidence interval.

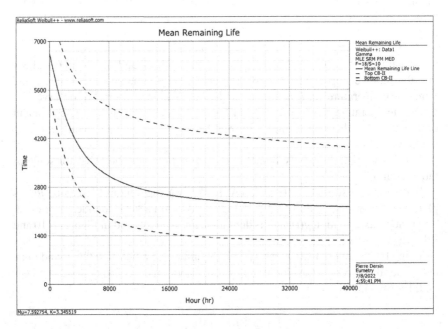

FIGURE 8.9   Mean residual life (MRL) for LED (I = 0.35 A, T = 85°C), with 90% confidence interval.

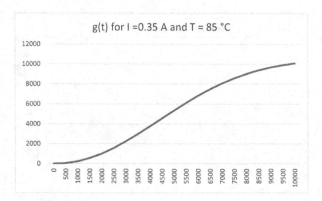

FIGURE 8.10   Transformation $g(t)$ for LED (I =0.35 A, T = 85°C).

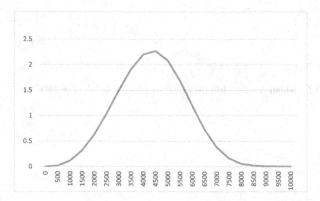

FIGURE 8.11   Metric dg/dt for LED (I = 0.35 A, T = 85°C).

For a higher current value (I = 0.7 A) and the same temperature (T = 85 °C), the two-parameter Weibull distribution is a very good fit, and the best estimates of the parameters are now η=4934h, β=3.85, resulting in an MTTF of only 4463h, down from 6826h or 6757h.

The k parameter estimate is now k=0.84, closer to 1, corresponding to a smaller standard deviation on the failure time.

The inflection point of the $g(t)$ curve—or the maximum of dg/dt, corresponds to t *= 2940h.

Thus, for t > 2940h, $\left| \dfrac{d\mathrm{MRL}}{dt} \right| \leq 0.84$.

The MRL curve is plotted in Figure 8.12 and the curves for $g(t)$ and dg/dt, in Figures 8.13 and 8.14, respectively. As could be expected, the MRL slope is steeper, and so is the rise of the time transformation curve $g(t)$. Thus the influence of the current intensity stress is indeed important.

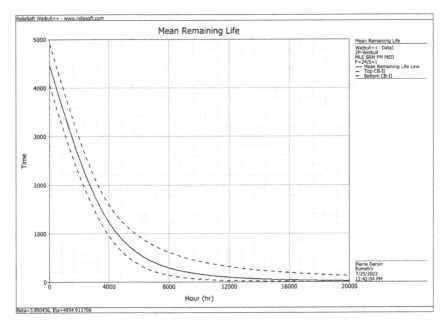

FIGURE 8.12   Mean residual life (MRL) for LED ($I = 0.7$ A, $T = 85$°C).

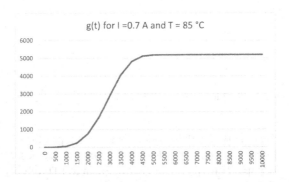

FIGURE 8.13   Transformation $g(t)$ for LED (I=0.7 A, $T = 85$°C).

From the knowledge of the g transformation, it is possible to determine confidence intervals for the RUL, using Theorem 5.2 (Chapter 5).

8.3.3   SECOND CASE STUDY: REDUNDANT SYSTEM

Now another use case is considered, which arises from the "ARAMIS challenge: Prognostics and Health Management in Evolving Environments" proposed at the European Safety & Reliability Conference in 2020 (ESREL), see Cannarile (2020) and Rocchetta (2022).

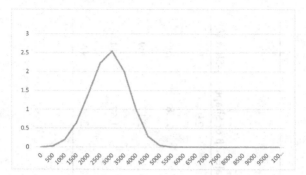

FIGURE 8.14   Metric dg/dt for LED (I = 0/7 A, T= 85°C).

It is a system made up of four parallel branches in active redundancy. Multiple sensors monitor and collect operational data at the component level and system level.

The data are acquired from a fleet of 200 identical systems and constitute the training set. The table in Appendix summarizes the available information.

In that table, each data row corresponds to a unit under test, and each column indicates the nature of the unit: one of four components, or the complete system. For each component, each entry of the table corresponds to either the first time an abnormal state is reached (as defined from the sensor measurements), or the end of the observations, as the case may be. For the system data, in the last column, the table entry indicates either the system failure time or the end of observations, i.e., the so-called mission time. Note that the degradations of the four components are not stochastically independent: some component degradations, or changes in component operational conditions, provoke degradations in others. The details can be found in the Aramis Challenge (ARAMIS Challenge 2020). See ARAMIS web site listed in the bibliography at the end of this chapter.

Using the same method as previously, the best probability distributions have been fitted to the data in order to estimate MTTF and other parameters at both component and system level. The MLE method is used as previously.

The results of the analysis are summarized in Table 8.8.

It is noticed that, for all four components, the k value is very close to 1 or, equivalently, the coefficient of variation is very small. Thus the failure times, or more precisely the times when the given degradation threshold is crossed, are almost deterministic.

Then the system has an MTTF (equal to 1098h) which is just marginally better than that of the most reliable component (1077h for Component 3).

This is an indication that the failures of the four components are far from being independent, otherwise the MTTF of the system, a "one-out-of four" configuration, would be a lot higher.

In fact, it is known that all components are impacted by a common cause, so that the redundancy is not very effective.

TABLE 8.8
Results of Field Data Analysis for "Aramis Challenge"

	μ/k (h)	MTTF (h)	k	CV	Inflection Point (h)	Distribution	η or Scale	β or Shape	γ-Shift (h)
Component 1	995	983	0.9878	0.078	913	3P Weibull	231	2.9	777
Component 2	1003	987	0.984	0.089	925	3P Weibull	216	2.28	796
Component 3	1114	1077	0.967	0.1295	985	3P Weibull	326	2.17	789
Component 4	1031	1003	0.973	0.1170	952	3P Weibull	262	1.67	814
System	1109	1098	0.9905	0.0692		Gamma	1.657	209	NA

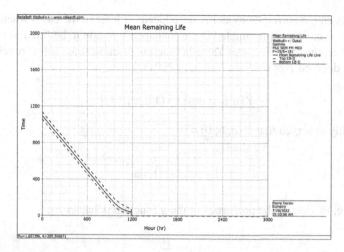

FIGURE 8.15 MRL estimate for "Aramis Challenge" system.

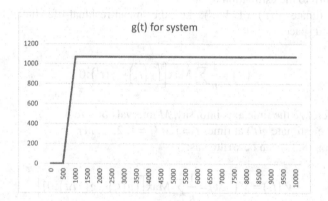

FIGURE 8.16 Time transformation g for "Aramis challenge" system.

As expected the coefficient of variation of the TTF at system level is smaller than for any of the components, or equivalently the k value is higher.

Accordingly, since that coefficient approaches unity, the MRL is almost a linear function of time, with slope close to -1, as shown in Figure 8.15, and the time transformation g (Figure 8.16) is almost a step function.

8.4 SURROGATE MODEL FOR THE $g(T)$ TRANSFORMATION

In Section 8.3, the estimate of the $g(.)$ function was derived from that of the reliability function, after estimation of the parameters μ and k.

A more straightforward approach would be to estimate the $g(.)$ function directly.

To this end, a surrogate model is introduced: at least in the cases where $g(t)$ is S-shaped, it can be approximated by an S-shaped function to be fitted to the data; for instance, the two-parameter Yamada function (used in some software reliability models), denoted $Y(a; b, t)$; see Yamada (1991).

$$Y(a,b;t) = a\left[1 - (1 + bt)e^{-bt}\right]. \tag{8.34}$$

It is readily observed that $Y(a, b; 0) = 0$
and

$$\lim_{t \to \infty} Y(a,b;t) = a. \tag{8.35}$$

Therefore, in view of the properties of the g function (Chapter 5),

$$a = \frac{\mu}{k}. \tag{8.36}$$

We now turn to the estimation of b.

The estimate $v'(t')$ of $V'(t')$, i.e. the mean-residual-life function in the transformed space, is:

$$v'(t') = \frac{1}{n}\sum_{j=1}^{j=n} \text{Max}\left[\left(g(t_j) - g(t');0\right)\right]. \tag{8.37}$$

Let us discretize the time axis into, say, M intervals of size Δt.

Thus we estimate $g(t')$ at times $t' = j\,\Delta t$ $(j = 1, 2, \dots M)$.

Equation (8.37) can be written as:

$$V'(t') = V'(j\Delta t) = \frac{1}{n}\sum_{i=1}^{i=n} \text{Max}\left[\left(g(t_i) - g(j\,\Delta t);0\right)\right] \tag{8.38}$$

$j = 1, 2, \dots M$.

By definition of the transformation g, $V(t')$ is a linear function of transformed time t'; therefore, when replacing the $g(t)$ function with the Yamada surrogate in (8.9), the parameters a and b will have to be selected in such a way that v' approximates a linear function of Δt.

An example is shown in Figure 8.17 where the parameter b has been found by trial and error with the numbers of Section 8.2. The best fitting value of b in this example is

$$\hat{b} = 8 \cdot 10^{-3} \,/\, \text{h}.$$

Then a is obtained from (8.36) and from the empirical estimations m and s of μ and σ, by applying (8.5) to (8.7).

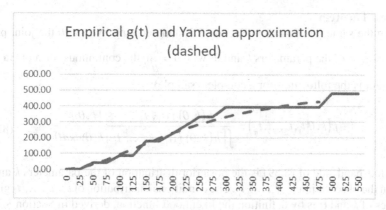

FIGURE 8.17 Empirical estimation $\hat{g}(t)$ and Yamada approximation.

The limitation of this method is that it does not provide a systematic way to estimate the coefficient b.

8.5 BAYESIAN ESTIMATION

As mentioned in Section 8.3, the knowledge from experts must always be accounted for and integrated with monitoring observation or test data. In frequentist statistics—based on the frequency of observed events, this expert knowledge can be exploited for model selection, either by selecting one out of several possible models or one out of several possible values of unknown parameters for a given model. The parameters to be estimated are numbers, unknown but assumed fixed, and deterministic. The frequentist approach is justified by the law of large numbers, which states that the frequency converges asymptotically to the probability as the sample size goes to infinity (cf. Birolini 2017). Instead, Bayesian statistics (see e.g. Berger 1988, Modarres 2016), uses the a priori knowledge of experts explicitly as an input in interpreting the observation data, and the parameters to be estimated are no longer defined as fixed but unknown numbers but as random variables following an a priori (or prior) distribution. Once observations are collected, that prior distribution is modified/updated into a posterior distribution using the Bayes theorem. Note that the posterior distribution merges both expert knowledge and observations and can be used as the new prior for the next round of observations, and so on. Bayesian methods applied to data sequences allow for a continuous or periodic update of the estimated model parameter. Bayesian methods are very useful in the context of condition-based monitoring where, thanks to the Internet of Things, physical factors related to the health state of the connected objects can be continuously measured, and RUL predictions updated accordingly. The Bayesian approach is also useful when events are rare and the associated data sets are relatively small. Since it provides a probability distribution for the parameters, the determination of a confidence interval is straightforward. The Bayesian estimation method is now briefly illustrated with the special distribution family of Chapter 3.

Bayes' Theorem

Using the same notation as in Section 8.2, and denoting by $h(k,\theta)$ the joint prior distribution of the parameters k and θ (with $\theta = \dfrac{k}{\mu}$), the continuous form of Bayes' theorem is the following (for a complete sample):

$$f\left(k,\theta|t_1,t_2,\ldots,t_n\right) = \frac{h(k,\theta)f(t_1, t_2,\ldots, t_n \mid k,\theta)}{\iint h(k,\theta)f(t_1, t_2,\ldots, t_n \mid k,\theta)dkd\theta}. \qquad (8.39)$$

The left-hand side of (8.39) is the posterior distribution of the parameters k and θ given the observations $t_1, t_2, \ldots, t_n$. The conditional distribution of $t_1, t_2,\ldots, t_n$ given values of k and θ is by definition the likelihood function, derived in Section 8.3:

$$f\left(t_1,t_2,\ldots,t_n \mid k,\theta\right) = L\left(t_1,t_2,\ldots,t_n \mid k,\theta\right). \qquad (8.40)$$

Thus L is given by (8.26). In the denominator of (8.39), the integration is carried out over the allowable range of the variables k and θ, namely:

$$0 \le k \le 1$$

$$\theta \le \frac{1}{\mathrm{Max}\left(t_i\right)}$$

Therefore the posterior distribution of (k, θ) is obtained as follows:

$$f\left(k,\theta|t_1,t_2,\ldots,t_n\right) = \frac{h(k,\theta)\mathrm{L}(t_1,t_2,\ldots,t_n \mid ;k,\theta)}{D}, \qquad (8.41)$$

where the denominator D is equal to:

$$D = \int_0^{\frac{1}{\mathrm{Max}(t_i)}} \theta^n \left[\int_0^1 \left(\frac{1}{k}-1\right)^n \prod_{i=1}^{i=n}\left(1-\theta t_i\right)^{\frac{1}{k}-2} h_1(k)dk \right] h_2(\theta)d\theta. \qquad (8.42)$$

It has been assumed that θ and k are independent random variables and therefore their joint distribution is the product of the marginal distributions:

$$h(k,\theta) = h_1(k)h_2(\theta). \qquad (8.43)$$

This is a reasonable assumption since, given the definition of θ, it is equivalent to assuming that k and μ are independent.

Choice of Prior Distributions

For instance, one could choose a gamma prior distribution for θ and a beta prior distribution for k (as this latter parameter takes values between 0 and 1).

Or, if no real prior knowledge is available for k, a uniform distribution over the interval $(0,1)$.

8.6 EXERCISES

1. Compute the estimates of $g(t)$ and $v'(t')$ according to the method of moments of Section 8.1 but using only the first five data points. Compare with the results of Section 8.1. What do you observe?
2. Show that s/m, in Section 8.1, is not an unbiased estimate of the coefficient of variation.
3. In Table 8.3, find other local maxima for the log-likelihood function. Why are there several local maxima? Build a simple code (for instance using Python and Matplotlib) to visualize the log-likelihood surface from several different angles.
4. For the example of Section 8.2, i.e., the likelihood function given by (8.23), derive the Fisher information matrix. Show how a confidence interval can be obtained for the parameters k and θ.
5. Consider the example of 8.2 but with censored data: the observations stop after 500h and the 5th unit has not failed yet. Generate the log-likelihood function and find the MLE estimate of k and θ.
6. Derive the formulas for Bayesian estimates of k and θ under the following prior distributions for the parameters: exponential distribution for θ and uniform distribution over $(0,1)$ for k.
7. Apply the results of Exercise 6 to the example of Section 8.2 and compare the results with those obtained with the MLE estimation method.

BIBLIOGRAPHY

AI Twilight, AI powered Digital twin for lighting infrastructure in the context of front-end Industry 4.0 Horizon 2020 EU Program (H2020-EU.2.1.1). See: https://github.com/Roberock/AI-TWILIGHT-TUe

ARAMIS Challenge 2020: The ARAMIS Challenge on Degradation State Assessment in evolving Environments.

ARAMIS Web site: https:// Innovation Challenges – Aramis srl (aramis3d.com)

Berger, J.O., *Statistical Decision Theory and Bayesian Analysis*. New York, Springer, 1988.

Birolini, A., *Reliability Engineering: Theory & Practice*. New York, Springer, 8th Ed., 2017.

Black, J.R., "Mass Transport of Aluminum by Momentum Exchange with Conducting Electrons," 6th Annual Reliability Physics Symposium (IEEE), 1967, pp. 148–159, doi:10.1109/IRPS.1967.362408

Cannarile, F., Compare, M., Baraldi, P., Yang, Z., & Zio, E., "Prognostics & Health Management in evolving Environments", *Proc. ESREL* 2020, 1085–1090, 2020.

Cinlar E., Bazani Z.P., & Osman E., "Stochastic Processes for Extrapolating Concrete Creep", *Journal of Engineering Mechanics Division*, 103(EM6), 1069–1088, 1977.

Di Bucchianico, A. & Castro, R., "Sequential and Nonparametric Statistics", Lecture Notes, Technische Universiteit Eindhoven, 2022.

Fisher, R.A., "On the Mathematical Foundations of Theoretical Statistics", *Philosophical Transactions of the Royal Society of London, Series A*, 222, 594–604, 1922.

Kotz, S., Johnson, N.L., Eds (1992), *Breakthroughs in Statistics*, vol. I, New York, Springer-Verlag, pp. 610–624.

Lei Ye, Xuezhi Gu, Dingbao Wang, & Richard M. Vogel, *An Unbiased Estimator of Coefficient of Variation of Streamflow*, Journal of Hydrology, Volume 594, Amsterdam, Elsevier, March 2021.

Luenberger, D.G., *Optimization by Vector Space Methods*. New York, J. Wiley & Sons, 1968.

Meeker, W.Q., Escobar, L.G., & Pascual, F.C., *Statistical Methods for Reliability Data*. Hoboken, NJ, J. Wiley & Sons, Second Edition, 2021.

Modarres, M., Kaminsky, M.P., & Krivtsov, V.V., *Reliability Engineering and Risk Analysis*, 3rd edition, Taylor & Francis, London, 2016. ISBN: 978-1498745871.

Pham, H., *Statistical Reliability Engineering: Methods, Models and Applications*. New York, Springer Nature, Technology & Engineering, 2021.

Raiffa, H., & Schlaifer, R., *Applied Statistical Decision Theory*. Hoboken, NJ, Wiley, 2000.

Rocchetta, R., Petkovic, M., Gao, Q., & Dimitrios, M., "A Robust Model Selection Framework for Fault Detection and System Health Monitoring with Limited Failure Examples: Heterogeneous Data Fusion and Formal Sensitivity Bounds", *Engineering Applications of Artificial Intelligence*, Vol. 114, 105140, 2022 https://doi.org/10.1016/j.engappai.2022.105140

Rocchetta, R., Zhao, Z., & Di Bucchianico, A., "An Interval Regression Model for High-Voltage LEDs Luminous Flux Maintenance Lifetime Estimation", ESREL, 2022.

van Noortwijk, J.M., "A Survey of the Application of Gamma Processes in Maintenance", *Reliability Engineering & System Safety*, 9, 2–21, 2009.

Yamada, S., Hishitani, J., & Osaki, S., "A Software Reliability Growth Model for Test-Effort Management," [1991] *Proceedings the Fifteenth Annual International Computer Software & Applications Conference*, 1991, pp. 585–590, doi: 10.1109/CMPSAC.1991.170243

APPENDIX: ARAMIS DATA CHALLENGE

Index	Component 1 TTF	Component 2 TTF	Component 3 TTF	Component 4 TTF	System TTF
1	945	933	1000	982	1000
2	1000	975	954	1000	1000
3	1000	1000	884	912	1000
4	1000	1000	1000	1000	1000
5	1000	858	917	1000	1000
6	949	984	1000	1000	1000
7	1000	1000	1000	1000	1000
8	970	843	1000	1000	1000
9	1000	821	929	839	1000
10	911	894	1000	917	964
11	935	1000	1000	903	1000
12	983	865	927	1000	1000
13	945	940	1000	1000	1000
14	1000	988	1000	1000	1000
15	859	936	883	1000	1000
16	1000	1000	926	930	1000
17	1000	904	978	1000	1000
18	1000	1000	873	858	1000
19	1000	1000	1000	938	1000
20	1000	1000	1000	1000	1000
21	997	1000	1000	1000	1000
22	1000	946	1000	1000	1000
23	879	885	953	1000	973
24	935	1000	870	922	1000
25	837	971	1000	966	1000
26	1000	1000	1000	930	1000
27	813	925	1000	878	1000
28	976	1000	899	890	1000
29	1000	941	1000	977	1000
30	1000	920	924	1000	1000
31	887	1000	1000	1000	1000
32	943	1000	1000	914	1000
33	861	1000	948	952	1000
34	1000	840	932	996	1000
35	1000	1000	937	1000	1000
36	894	1000	860	938	1000
37	954	1000	1000	918	1000
38	1000	1000	1000	1000	1000
39	891	1000	1000	1000	1000
40	947	1000	879	907	986
41	1000	860	1000	821	1000
42	1000	986	909	893	1000
43	1000	996	1000	918	999

(Continued)

Index	Component 1 TTF	Component 2 TTF	Component 3 TTF	Component 4 TTF	System TTF
44	1000	893	1000	931	1000
45	939	965	1000	1000	1000
46	1000	900	916	958	1000
47	914	1000	801	921	1000
48	944	916	979	1000	1000
49	869	994	1000	1000	1000
50	1000	844	836	907	988
51	1000	843	980	1000	1000
52	935	1000	915	877	1000
53	1000	1000	942	834	1000
54	829	852	1000	856	940
55	852	1000	940	870	949
56	860	935	974	1000	1000
57	1000	1000	992	1000	1000
58	1000	1000	1000	897	1000
59	861	911	1000	1000	1000
60	888	1000	1000	983	1000
61	1000	971	894	929	1000
62	992	1000	841	930	1000
63	1000	895	997	1000	1000
64	985	1000	1000	875	1000
65	1000	1000	897	880	1000
66	930	1000	943	889	1000
67	1000	871	867	1000	1000
68	1000	896	911	878	933
69	930	1000	851	870	961
70	890	1000	1000	854	1000
71	952	1000	1000	1000	1000
72	939	1000	1000	921	1000
73	1000	938	992	1000	1000
74	944	989	864	1000	1000
75	993	1000	1000	963	1000
76	1000	907	1000	972	1000
77	841	1000	902	1000	1000
78	1000	906	859	1000	1000
79	896	972	1000	889	1000
80	906	1000	920	878	951
81	928	1000	1000	1000	1000
82	932	1000	904	889	1000
83	1000	926	865	1000	1000
84	982	997	1000	896	1000
85	887	1000	1000	907	1000
86	1000	912	958	1000	1000
87	1000	1000	939	945	1000
88	1000	935	1000	931	1000

Index	Component 1 TTF	Component 2 TTF	Component 3 TTF	Component 4 TTF	System TTF
89	978	1000	931	972	1000
90	1000	926	857	840	940
91	1000	921	869	880	1000
92	972	951	1000	884	984
93	1000	1000	1000	899	1000
94	868	948	1000	1000	1000
95	1000	1000	913	1000	1000
96	877	945	1000	1000	1000
97	981	1000	981	990	1000
98	924	973	1000	1000	1000
99	1000	993	1000	898	1000
100	1000	948	859	953	1000
101	1000	1000	1000	1000	1000
102	1000	1000	1000	1000	1000
103	971	915	1000	884	998
104	819	959	1000	1000	1000
105	997	889	947	1000	1000
106	1000	1000	979	974	1000
107	1000	927	940	1000	1000
108	897	1000	1000	870	1000
109	1000	883	1000	1000	1000
110	896	1000	1000	915	1000
111	989	1000	1000	853	1000
112	1000	1000	1000	1000	1000
113	873	880	906	1000	957
114	923	937	1000	1000	1000
115	978	954	1000	1000	1000
116	975	941	1000	1000	1000
117	968	945	1000	1000	1000
118	921	1000	960	905	1000
119	884	913	1000	1000	1000
120	916	1000	895	825	1000
121	1000	1000	927	854	1000
122	1000	1000	1000	917	1000
123	917	1000	965	1000	1000
124	890	1000	923	904	1000
125	946	1000	950	896	1000
126	1000	917	1000	953	1000
127	978	1000	939	1000	1000
128	976	1000	896	1000	1000
129	1000	1000	1000	1000	1000
130	918	938	925	1000	1000
131	966	1000	968	881	1000
132	1000	1000	1000	1000	1000
133	1000	1000	1000	882	1000

(*Continued*)

Index	Component 1 TTF	Component 2 TTF	Component 3 TTF	Component 4 TTF	System TTF
134	916	1000	1000	895	1000
135	1000	1000	1000	1000	1000
136	892	914	866	1000	974
137	1000	942	937	982	989
138	980	1000	912	886	1000
139	1000	1000	1000	1000	1000
140	1000	917	876	1000	1000
141	1000	948	940	1000	1000
142	882	993	930	1000	1000
143	905	871	1000	958	994
144	1000	985	880	902	1000
145	1000	827	898	821	1000
146	964	1000	945	1000	1000
147	914	1000	938	990	1000
148	1000	1000	956	986	1000
149	831	1000	986	902	988
150	944	899	1000	1000	1000
151	1000	1000	870	1000	1000
152	1000	1000	1000	840	1000
153	966	899	1000	1000	1000
154	926	956	933	1000	1000
155	1000	886	950	922	1000
156	913	930	1000	890	1000
157	1000	1000	1000	1000	1000
158	987	1000	1000	1000	1000
159	1000	881	933	1000	1000
160	922	1000	898	981	1000
161	1000	1000	885	1000	1000
162	991	1000	1000	883	1000
163	926	1000	947	1000	1000
164	1000	1000	857	1000	1000
165	1000	876	1000	899	1000
166	1000	1000	1000	1000	1000
167	904	919	902	1000	1000
168	949	912	1000	1000	1000
169	869	983	1000	1000	1000
170	1000	1000	849	1000	1000
171	1000	1000	981	1000	1000
172	934	1000	1000	1000	1000
173	1000	1000	967	982	1000
174	1000	963	891	1000	1000
175	1000	999	905	1000	1000
176	942	1000	907	903	1000
177	1000	846	1000	983	1000
178	1000	1000	1000	864	1000

Index	Component 1 TTF	Component 2 TTF	Component 3 TTF	Component 4 TTF	System TTF
179	1000	1000	1000	910	1000
180	1000	1000	1000	973	1000
181	930	1000	920	988	1000
182	1000	1000	922	961	1000
183	1000	1000	922	924	1000
184	945	942	1000	1000	1000
185	970	1000	914	925	1000
186	1000	946	970	958	1000
187	972	938	1000	1000	1000
188	922	944	1000	1000	1000
189	1000	861	1000	928	1000
190	816	830	1000	1000	896
191	1000	926	981	1000	1000
192	1000	1000	830	1000	1000
193	984	832	905	1000	1000
194	843	845	1000	1000	1000
195	1000	965	1000	944	1000
196	839	1000	823	897	1000
197	949	1000	903	890	1000
198	1000	965	1000	1000	1000
199	895	1000	1000	929	1000
200	1000	1000	1000	1000	1000

9 Implications for Maintenance Optimization

9.1 INTRODUCTION

In previous chapters, the subject of study has been the probabilistic characterization of remaining useful life (RUL), also known as the remaining time to failure. A general definition of failure has been introduced together with a mathematical framework for failure modeling and analyses. However, no consideration of repair, restoration, or any other action preventing a failure or remedying the effect of a failure, has been made. In other words, the systems and components were treated as "non-repairable" units.

The purpose of this chapter is to outline ways in which the study of the RUL and its time dynamics can support maintenance decisions. After a brief reminder of the notion of maintenance, maintenance costs, and risks associated with maintenance decisions, this chapter addresses the mathematical modeling of these costs and risks.

Methods are then presented for optimizing costs, first by selecting the periodicity of scheduled preventive maintenance, and then by applying predictive maintenance.

The literature on this subject is extremely abundant and still growing (see for instance Campbell 2022, in particular Chapter 12 by Jardine, on optimization of preventive maintenance periodicity, such as by block replacement and age replacement), and it is not meant here to duplicate the excellent coverage already existing, but rather to present a viewpoint consistent with the results presented in the previous chapters, which, it is felt, can bring some added value.

In Section 9.2, a reminder of the different possible maintenance strategies is presented, with a discussion of the trade-offs between cost and risk.

Section 9.3 addresses classical methods for quantifying those costs and risks in the context of traditional systematic preventive maintenance; with particular application to the distributions with MRL linear in time.

Finally, in Section 9.4, a conceptual framework is presented for cost optimization with a predictive maintenance strategy, which implies a dynamic maintenance policy, i.e., dynamic updating of the preventive maintenance times based on evolving asset condition, with a failure risk which can be bounded a priori.

DOI: 10.1201/9781003250685-9

9.2   MAINTENANCE DECISIONS: BALANCING COSTS AND RISKS

Maintenance is defined (IEC 60030-6-14) as "the combination of all technical and administrative actions intended to retain an item in, or to restore it to, a state in which it can perform a required function".

By definition, preventive maintenance takes place before any failure has occurred, and is meant to prevent failures from happening; while corrective maintenance takes place after a failure, i.e., once a function has been lost, and is aimed at making the item operational again, that is, restoring the function. This can be achieved by repairing a failed item of equipment, or by replacing the failed item with a healthy one.

A key issue in deciding on a maintenance strategy is whether to perform some preventive maintenance and, if so, what type of preventive maintenance rule should be followed.

The drawback of corrective maintenance is that it is unscheduled: by definition, the corresponding maintenance operations cannot be planned, since they are triggered by random events: the failures. Therefore the corresponding logistics (arranging for spare parts and specialized maintenance personnel to be available to perform the needed maintenance operation) tend to be expensive, especially under emergency conditions. In addition, in many contexts, those failures have a significant impact on operations—such as electric power unscheduled outages, canceled flights, train delays, etc. This is true at least for complex systems with important operational impacts. From a business contractual viewpoint, such systems are often sold with a service-level agreement, i.e., the supplier commits to a level of service which is measured by an availability level or similar metrics, and he has to pay important penalties if those contractual objectives are not met.

In contrast, for consumables, such as electric lights, with no service level agreements, suppliers have no particular incentive to avoid failures and even see the provision of spare parts as a welcome additional source of revenue. In those cases, preventive maintenance is not so easy to justify even if it is technically feasible. An example of such a situation is provided, for instance, by nozzles in plasma cutting or plasma welding torches.

Preventive maintenance (Figure 9.1) can be either scheduled or condition-based. The former, also known as systematic or planned maintenance, is defined by maintenance intervals which are predetermined and can be measured in operating time, mileage, cycles, or in any other usage variable which can be seen as a proxy variable for aging. In contrast, in "condition-based maintenance" (CBM), actions and decisions are informed by estimates of the current condition of the asset to be maintained, e.g., its health and degradation state.

Beyond maintenance cost reduction, the intended benefits of CBM include an increased asset utilization rate and, in some cases, a reduction in maintenance personnel exposure to hazardous conditions (think for instance of night inspections of assets such as railway point machines, or inspections in nuclear power plants).

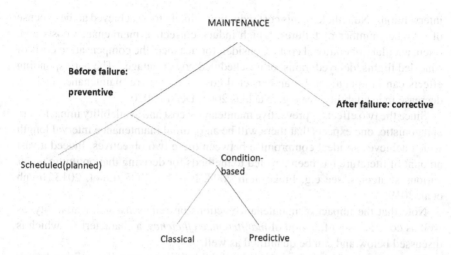

FIGURE 9.1 Different types of maintenance policies.

Usage variables are relevant for both systematic and CBM, and an appropriate definition is application-specific. For instance, the key usage variable of electrical equipment is often the power-on time, while for mobile mechanical equipment, such as wheels, cranes, and cars, the mileage is more relevant. Finally, for equipment which is cyclically operated, such as train passenger access doors, the number of on-off (opening-closing) cycles is the most relevant indicator of aging.

In scheduled preventive maintenance, the decision to be made then is to choose the length of the maintenance interval: after how much time, how much mileage or how many cycles should preventive maintenance take place?

In determining the maintenance interval, two main criteria must be considered: the maintenance and operations costs on the one hand, and availability or service reliability (frequency of service-affecting failures) on the other hand.

The cost function must be clearly defined, and its scope depends on the viewpoint adopted: are we trying to optimize the total cost to the operator, or the cost to an original equipment manufacturer (OEM) with or without a service level agreement, or the cost to a maintenance provider?

The costs may include the labor, materials, operational impact of failures, and other costs induced by preventive maintenance (including sometimes indirect costs such as spare parts holding costs) and by corrective maintenance (such as unscheduled maintenance operations after failures). On the other hand, availability is measured by a figure of merit, which can be the classical availability (the percentage of time the asset is available to perform its function) or, in the case of systems, a quality-of-service measure such as the percentage of planned missions (train trips, scheduled flights) which are actually fulfilled.

It is intuitively evident that longer maintenance intervals lead to lower the cost of preventive maintenance actions due to less frequent maintenance crew interventions. Nonetheless, this cost reduction is likely to be achieved at the expense of a higher number of failures, which induce corrective maintenance costs and, often, ancillary operational costs. Consider, for instance, the compensation costs of canceled flights, delayed trains, or unscheduled power outages. The corresponding effects can be quantified by an expected cost of corrective maintenance or by a decrease in availability, see e.g., Nachlas 2005, Dersin 2020.

Since the two effects—preventive maintenance cost and availability impact—are antagonistic, one expects that there will be an optimal maintenance interval length which achieves an ideal compromise between these two objectives. Indeed a vast amount of literature has been devoted to methods for deriving that optimum under various strategies, see e.g. Finkelstein 2008, Nachlas 2005, Letot, 2013,Huynh et al. 2019.

Note that the impact of maintenance actions on reliability and availability, as well as cost, is also a function of *maintenance efficiency,* a characteristic which is discussed below and can be quantified as well.

The same considerations apply, in a slightly different way, if condition-based, rather than scheduled, maintenance is applied. CBM consists of basing the maintenance decisions on the current condition, health, or degradation state of an asset to be maintained. That condition, or state, is assessed either periodically, or continuously through condition monitoring. Some criterion, such as a health state indicator, is generally used to trigger a maintenance operation if the asset's condition is deemed unacceptably degraded. For instance, in the case of passenger train access doors, the pattern of the current curve versus time corresponding to a door opening-closing cycle can be used to build a health indicator (Dersin & Thévenet 2019); or a similar current curve can be used for the CBM or railway switches (Alessi 2016).

One of the main differences between the cost assessment for scheduled maintenance and condition-based maintenance policies is that, for the former, the operational costs are deterministic since the frequency of maintenance actions is known a priori. In contrast, CBM actions and associated costs are uncertain and affected by randomness due to a non-deterministic deterioration of the health state of the monitored asset.

Predictive maintenance is the special CBM strategy which takes into account, not only the current asset condition, but also its likely evolution. Then the key indicator to support the maintenance decision is the estimated RUL, as the primary goal of the maintenance strategy will be to decide on the timing of a preventive maintenance action in such a way as to avoid the impending failure as much as possible.

As emphasized in previous chapters, any RUL estimation includes some amount of uncertainty; therefore maintenance decisions based on RUL necessarily include a risk: the risk of maintaining too late and thus not preventing the failure, which is to be balanced against the cost of additional preventive maintenance.

9.3 QUANTIFYING COSTS AND RISKS

9.3.1 REJUVENATION AND MAINTENANCE EFFICIENCY

One useful way to look at maintenance in connection with aging and MRL is to think of it in terms of "asset rejuvenation". The goal of a maintenance action can be viewed as bringing the asset back to an earlier state when it was fully healthy; its residual reliability is then the same as the initial reliability at the beginning of its life. Therefore its MRL is also equal to its initial value MRL(0)= MTTF (see Figure 9.2(a), where the RUL is shown to be restored to its initial value after a maintenance operation).

Then, after the maintenance action, the asset has been rejuvenated and starts a new life: it is said to be "as good as new" (AGAN)—(Nachlas 2005, Finkelstein 2008).

However, this is an ideal goal which often cannot be achieved, as it requires correcting all the ongoing degradation processes. Most of the time, the maintenance action remedies a particular issue but does not act on other ongoing degradations. To take an example from everyday life, assume an automobile is brought to the garage to replace a badly worn wheel; the servicemen look at the wheel only but have not paid attention to wear in the transmission; when leaving the workshop with a new wheel, the car will not be "as good as new" because the transmission has been wearing out since the beginning and the maintenance action has not fixed that issue. Only a thorough overhaul would have permitted inspecting and addressing all the issues.

At the other end of the spectrum from AGAN is "as bad as old" (ABAO), or "minimal repair" (Nachlas 2005, Finkelstein 2008), which is the state of an asset when the maintenance action that has been performed is minimal, i.e., just sufficient to allow it to continue to operate; then it restarts operation with the residual

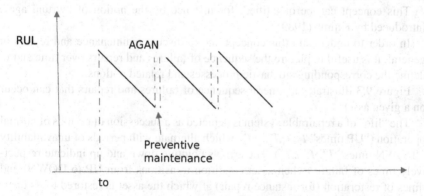

FIGURE 9.2(a) Successive operating periods (perfect maintenance: AGAN).

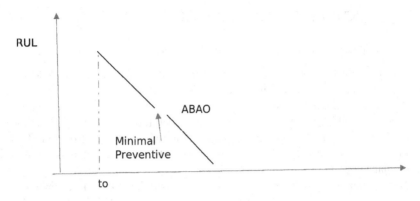

FIGURE 9.2(b)   Successive operating periods (minimal maintenance: ABAO).

reliability it had just before the maintenance action (Figure 9.2(b) shows such a case, where the asset has exactly the same RUL after the maintenance as it had just before).

Minimal repair is normally less costly than perfect repair because it includes fewer and simpler maintenance actions—just a superficial fix in fact; but, after a minimal repair, the next failure will occur earlier than after a perfect repair; therefore, in the long run, this strategy leads to higher maintenance costs. In fact there is a trade-off between paying now and paying later.

The general case lies between those two extremes: after the maintenance action, the asset has an MRL which is between the initial one (MTTF) and the one it had just before maintenance.

See, e.g., (Finkelstein 2015) for a discussion on the degree of optimal repair, i.e., what the optimal maintenance efficiency is for a given cost function and given reliability characteristics.

This concept can be quantified, for instance by the notion of "virtual age", introduced by Kijima (1989).

In order to understand this concept and to discuss maintenance and repairs in general, it is useful to picture the sequence of failures and repairs over time and to define the corresponding stochastic processes and related notions.

Figure 9.3 illustrates a general sequence of failures and repairs that can occur on a given asset.

The "life" of a repairable system is depicted as a succession of periods of normal operation ("UP times" $T_0, T_1, T_2, \ldots T_n$) which alternate with periods of unavailability ("DOWN times" $T_1', T_2' \ldots T_n'$). The arrows pointing down and up indicate respectively times of failure (when asset condition transitions from UP to DOWN) and times of restoration (for instance repair) at which the asset is restored to its operational state (from DOWN to UP). Restoration may occur without actual repair (for instance though a replacement).

For the present discussion it will be assumed that durations of the downtime periods are negligible as compared to those of the uptime periods: $T_i' \ll T_i$.

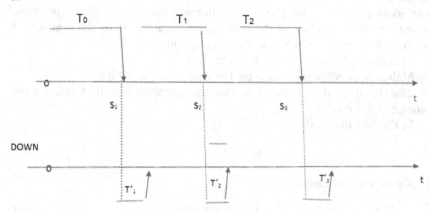

FIGURE 9.3 Successive failure and restoration times.

Let us also express the variables that correspond to the failure times. Denoting by S_n the time of occurrence of the nth failure,

$$S_n = \sum_{i=0}^{i=n-1} T_i. \tag{9.1}$$

The number of failures over a time interval of length t, $N(t)$, is related to those time periods in an obvious way (see also Figure 9.3):

$$N(t) = \begin{cases} 0, & t < T_0 \\ n, & S_n \le t < S_{n+1} \end{cases} \quad n = 1,2.. \tag{9.2}$$

so the events $[N(t) \ge n]$ and $[S_n \le t]$ are equivalent. In other words, it is equivalent to say that at least n failures have occurred up to time t, or to say that the time to the nth failure does not exceed t. The stochastic process $\{N(t)\}$ is called a counting process.

In the case of perfect repair (AGAN), the successive uptime periods are independent and identically distributed (iid) since, after each restoration, the new asset life starts in exactly the same conditions as if it were completely new.

Then the process $\{T_i\}$ is by definition a renewal process (Ross 2010), a type of stochastic process that enjoys many useful mathematical properties.

Instead, when the repair is minimal, the past history influences the asset's propensity to fail, as it is not at all renewed by the maintenance operations. Otherwise stated, successive operating periods do not have the same distribution. Therefore they cannot be modeled by a renewal process. Below will be described a more-adequate model for that case.

Now the notion of virtual age is a construct that measures how much age has been removed by the maintenance operation. Kijima in fact introduced two models .

In the first one, the maintenance operation is assumed to remove a fraction of the age accumulated since the previous maintenance operation; while in the second Kijima model, a fraction of the total age accumulated since the beginning is assumed to be removed by the maintenance operation.

Mathematically, they take the following form.

Maintenance efficiency is measured by a parameter a, $0 \le a \le 1$.

The virtual age A_i after the ith maintenance operation is defined with that parameter, as follows:

In Kijima's first model:

$$A_i = A_{i-1} + aT_i \tag{9.3}$$

In Kijima's second model:

$$A_i = a\,(A_{i-1} + T_i). \tag{9.4}$$

The AGAN case corresponds, in both models, to $a = 0$, so that the virtual age is always equal to 0; and the ABAO case, to $a = 1$, so that the virtual age corresponds to the real age.

The fraction of accumulated age which is removed by the maintenance action is $(1-a)$.

In Kijima Model I, by writing (9.3) as

$$A_i = A_{i-1} + T_i - (1-a)T_i,$$

it is apparent that the fraction $(1-a)$ of the age accumulated since the previous maintenance action is removed from virtual age; whereas, in Kijima Model II, from (9.4),

$$A_i = A_{i-1} + T_i - (1-a)(A_{i-1} + T_i),$$

which shows a fraction $(1-a)$ of the total accumulated virtual age is assumed to be removed by the maintenance action.

An alternative to Kijima's model is the Pham & Wang model (Pham 1996), where the successive intervals T_n of normal operation are modeled as

$$T_n = \alpha^n X_n$$

with $0 < \alpha < 1$ and the X_n random variables are iid. The α parameter captures the effect on lifetime of each maintenance operation.

More recently, Doyen and Gaudoin (2002) have provided comprehensive families of models that describe imperfect maintenance and fall into two classes: arithmetic reduction of age models (to which Kijima and Pham belong) and arithmetic reduction of intensity models where, after maintenance, the failure intensity is considered reduced to a variable extent according to maintenance efficiency.

We now first address the question of optimizing the periodicity of scheduled maintenance, before turning to predictive maintenance. Only the two extreme cases are considered in what follows: ABAO and AGAN; not the intermediate, partial repair cases. The time required for maintenance (corrective or preventive) is assumed to be negligible as compared to the normal operation periods.

9.3.2 MINIMAL MAINTENANCE (ABAO)

When repair is minimal (Figure 9.2b), the history of the asset before the failure will influence the time to the next failure. The number of failures over a time interval $(0,t)$, $N(t)$, increases with time, on average; otherwise stated, its mathematical expectation is a non-decreasing function of time.

Several mathematical models have been proposed to model $N(t)$. (See Finkelstein 2008 for a detailed mathematical discussion, and Nachlas 2005 for a more intuitive exposition.)

The most common one is the nonhomogeneous Poisson process. The stochastic process $\{N(t)\}, t \geq 0$, is called a Poisson process with intensity function $\lambda(t)$ (Ross 2010) if it satisfies the following properties:

i. $N(0) = 0$.
ii. It has independent increments, i.e., the numbers of events that occur in disjoint intervals are stochastically independent.
iii Rare events: in a short interval of time, there occurs at most one event. This is formalized by stating that, in an interval of length h, the probability of more than one event is an $o(h)$, a term of higher order in h:

$$P\left[N(t+h) - N(h) > 1\right] = o(h) \qquad (9.5)$$

with $o(h)$ the usual notation for a quantity of higher order in h:

$$\lim_{h \to 0} \frac{o(h)}{h} = 0.$$

Then, by definition of the intensity function $m(t)$,

$$P\left[N(t+h) - N(t) = 1\right] = m(t)h + o(h). \qquad (9.6)$$

In general, the intensity function $m(t)$ varies with time, and the process is called a nonhomogeneous Poisson process (NHPP).

A special case is the homogeneous Poisson process (HPP), where the intensity $m(t)$ is constant with time: $m(t) = \lambda$.

Note that the notion of intensity function is in general different from that of failure rate: the failure rate is a conditional probability per unit of time (Chapter 2), while the intensity is the time derivative of the expected number of failures.

The expected number of failures over a time interval $(0,t)$ for a Poisson process is given by the integral of the intensity function:

$$E[N(t)] \equiv M(t) = \int_0^t m(s)\,ds. \tag{9.7}$$

Only in the case of a homogeneous Poisson process do failure rate and failure intensity coincide, and then:

$$E[N(t)] = \lambda t. \tag{9.8}$$

Returning to the process of successive failures (Figure 9.3) with minimal repairs, the assumptions that characterize the NHPP model are considered adequate (cf. Finkelstein 2008). The intensity function must be consistent with the failure rate of the first life T_0, which is the failure rate corresponding to the TTF distribution.

Thus the $\{N(t)\}$ process is modeled with an intensity function $\lambda\,(t)$, and

$$E[N(t)] = \int_0^t \lambda(s)\,ds = \Lambda(t), \tag{9.9}$$

the cumulative hazard function.

Therefore the probability distribution of the number $N(t)$ of failures over $(0,t)$ is a Poisson distribution:

$$P[N(t) = n] = e^{-\Lambda(t)} \frac{(\Lambda(t))^n}{n!} \tag{9.10}$$

As shown in Chapter 3, in the special family of distributions with an MRL linear in time, the cumulative hazard function is

$$\Lambda(t) = -\frac{(1-k)}{k} \log(1 - \frac{kt}{\mu}), \tag{9.11}$$

a decreasing function of k.

Therefore, the higher k is (the closer to 1), the smaller the average number of failures over a given time interval $(0,t)$.

The two limiting cases are, as usual, $k=0$ (exponential time to failure), which results in

$$\Lambda(t) = \frac{t}{\mu}, \tag{9.12}$$

i.e., a constant failure intensity equal to $\lambda = \dfrac{1}{\mu}$, and $k = 1$ ("Dirac", deterministic distribution), which leads to

$$\Lambda(t) = 0 \quad \text{for } t < \mu.$$

9.3.3 PERFECT MAINTENANCE (AGAN)

Consider now the case when preventive maintenance takes place with periodicity T, and is perfect, so that the asset is AGAN after a maintenance operation (corrective or preventive).

Then the times between successive failures, $T_1, T_2, ..., T_n$ constitute a sequence of independent, identically distributed, nonnegative random variables. The corresponding counting process, $\{N(t)\}$, is then a renewal process (Ross 2010).

The renewal function, $H(t)$, is the mathematical expectation of the number of renewals,

$$H(t) = E[N(t)] \tag{9.13}$$

which is what is denoted $M(t)$ in Section 9.3.3.

Its derivative,

$$h(t) = \frac{dH}{dt} \tag{9.14}$$

satisfies the following convolution equation, called the renewal equation:

$$h(t) = f(t) + \int_0^t h(x) f(t-x) dx \tag{9.15}$$

(Birolini 2017)

This is easily proved as follows.

From the equivalence of the events $[N(t) \geq n]$ and $[S_n \leq t]$, or $[N(t) < n]$ and $[S_n > t]$, there follows:

$$P[N(t) = n] = P[N(t) \leq n] - P[N(t) \leq n-1] = P[S_n \leq t] - P[S_{n-1} \leq t] \tag{9.16}$$

Or, denoting by $F_n(t)$ the cumulative distribution function of S_n—defined by (9.1),

$$P[N(t) = n] = F_n(t) - F_{n+1}(t). \tag{9.17}$$

Therefore

$$H(t) = E[N(t)] = \sum_{n=1}^{\infty} nP[N(t) = n] = \sum_{n=1}^{\infty} n[F_n(t) - F_{n+1}(t)] = \sum_{n=1}^{\infty} F_n(t). \tag{9.18}$$

The probability density functions f_n of the variables S_n satisfy the following relation (by conditioning upon the time of occurrence of the nth failure):

$$f_n(t) = \int_0^t f(x) f_{n-1}(t-x) dx. \tag{9.19}$$

In other words, $f_n(t)$ is the convolution of $f(t)$ and $f_{n-1}(t)$:

$$f_n = f * f_{n-1}$$

By changing the order of summation and integration, the renewal density $h(t)$ is obtained from (9.18):

$$h(t) = \sum_{n=1}^{\infty} f_n(t). \tag{9.20}$$

Then the renewal equation (9.15) follows from (9.19) and (9.20)

By taking the Laplace transforms of both sides (denoting by $\widetilde{h}(s)$ the transform of $h(t)$), the renewal equation (9.15) is equivalent to:

$$\tilde{h} = \tilde{f} + \tilde{h}.\tilde{f} \tag{9.21}$$

Or

$$\tilde{h} = \frac{\tilde{f}}{1 - \tilde{f}} \tag{9.22}$$

and $h(t)$ can be obtained by inverting the Laplace transform $\tilde{h}(s)$, then $H(t)$ is obtained by integration of $h(t)$. For a fleet of identical assets (such as aircraft engines, train wheels), one distinguishes "block replacement" and "age replacement" policies (see e.g. Nachlas 2005).

In the block replacement policy, all identical assets in a fleet are replaced at some point in time, regardless of their age.

In the age replacement policy, each device is replaced when it reaches a predetermined age.

Here we briefly present a well-known cost model for the block replacement policy (age replacement policy is also examined in Nachlas 2005, for instance).

Denote by c_1 the cost of a planned replacement and by c_2 the cost of a failure (which is not just the cost of a corrective maintenance operation, but the total cost induced by the failure, which may include loss of production, cost of operation disruption, etc.). Then, if the periodicity of replacements is denoted T, the average cost per unit of time is obtained as

$$E\left(\frac{\text{Cost}}{T}\right) = \frac{c_1 + c_2 H(T)}{T}. \tag{9.23}$$

This approach is valid only when corrective maintenance is perfect (AGAN), then the successive failure times constitute a renewal process (Birolini 2017, Nachlas 2005).

The optimal replacement period T^* is the one that minimizes the cost per unit of time; it can be obtained by setting to zero the derivative of the right-hand side of (9.23), which yields

$$Th(t) - H(t) = \frac{c_1}{c_2}. \tag{9.24}$$

It is seen that the solution depends only on the ratio between the two costs c_1 and c_2. In the special case of a constant failure rate, $H(t) = \lambda t$ and $h(t) = \lambda$, so that (9.24) has no solution. This is consistent with the fact that, in the case of a constant failure rate (exponential time-to-failure distribution), preventive maintenance is useless.

For the special family of distributions studied in Chapter 3, an expression for the Laplace transform has been given:

$$\tilde{f}(s) = \sum_{n=0}^{n=\infty} (-1)^n \frac{(\mu s)^n}{k^n} \frac{\Gamma\left(\frac{1}{k}\right)}{\Gamma\left(n + \frac{1}{k}\right)}. \tag{9.25}$$

For a general distribution, the optimal solution T^* can be found in the $g(T)$ space, and then converted back in the original time by the transformation g^{-1}.

Consider for instance the special case of a uniform distribution ($k = \frac{1}{2}$) over the interval $(0, 2\mu)$. Then (9.25) becomes

$$\tilde{f}(s) = \sum_{n=0}^{n=\infty} (-1)^n \frac{(2\mu s)^n}{(n+1)!} = \frac{1 - e^{-2\mu s}}{2\mu s}. \tag{9.26}$$

For the special case of a "Dirac distribution" ($k = 1$),

$$\tilde{f}(s) = \sum_{n=0}^{n=\infty} (-1)^n \frac{(\mu s)^n}{n!} = e^{-\mu s}. \tag{9.27}$$

Therefore

$$\tilde{h}(s) = \frac{e^{-\mu s}}{1 - e^{-\mu s}} = \frac{1}{e^{\mu s} - 1}. \tag{9.28}$$

Accordingly (see, e.g., Selby 1972), $H(t)$ is the staircase function defined by

$$H(t) = \begin{cases} 0 \text{ if } 0 \le t < \mu \\ n\mu \le t < (n+1)\mu \end{cases}. \tag{9.29}$$

In other words,

$$H(t) = Int\left(\frac{t}{\mu}\right),$$

the largest integer not exceeding $\frac{t}{\mu}$.

which is intuitive since, in the Dirac distribution case, failures occur deterministically at times that are multiples of μ.

Finally, in the exponential distribution case ($k=0$), one finds, by taking the limit k→0,

$$\tilde{f}(s) = \frac{1}{1 + \mu s} \qquad (9.30)$$

whence it follows that

$$\tilde{h}(s) = \frac{1}{\mu s}$$

and therefore,

$$h(t) = \begin{cases} \dfrac{1}{\mu} & for\, t > 0 \\ 0 & for\ t \leq 0 \end{cases} \quad (step\,function)$$

and

$$H(t) = \frac{t}{\mu} \quad for\ t > 0 \qquad (9.31)$$

which corresponds to a homogeneous Poisson process of rate $\lambda = \dfrac{1}{\mu}$ for the number of failures in a time interval $(0, t)$.

We now turn to predictive maintenance.

9.4 PREDICTIVE MAINTENANCE

The real benefit of predictive maintenance, which sets it apart from the traditional approach of scheduled maintenance, results from the fact that maintenance times can be continuously adjusted to reflect the evolving estimation of the asset's RUL. As new data is acquired and transmitted, the RUL must be updated. RUL estimates include a degree of uncertainty due to measurement inaccuracy, model imprecision, imperfect knowledge, the stochastic nature of the degradation processes, and the unknown future mission profiles. It is, therefore, necessary to present the RUL estimates with confidence intervals, which quantify the level of imprecision affecting the RUL predictions and will result in wider bounds the more severe the uncertainty and the more sources of uncertainty are accounted for. Note that RUL estimates and confidence intervals that can be useful for predictive maintenance must always be of an order of magnitude compatible with the logistic constraints of the maintenance decision-making process. For instance, if maintenance crew needs 8h to reach the premises where the asset is located, an RUL estimate of 1h is useless. Also, the accuracy of the confidence interval must be proportionate to the application context.

Figure 9.4 illustrates the lower and upper bound of the RUL confidence interval in the case of a distribution with MRL linear in time, i.e. in the $g(T)$ space.

In the particular example, it is assumed $k = 0.9$. The uncertainties which are modeled in this way relate only to the stochasticity of the degradation process. If measurement uncertainty is to be taken into account as well, then k must be estimated by one of the methods of Chapter 8, with a confidence interval.

Actually Figure 9.4 is an adaptation of Figure 3.4 of Chapter 3; here, it is presented differently: on the x-axis is time and, on the y-axis, the 80% confidence interval for the EOL (end of life), i.e. the sum of current time and RUL.

For instance, at $t = 40$h, the RUL is estimated to be between 43h (s-) and 71h (s+), which therefore implies a maintenance intervention at a time (t') between 83h and 111h.

A risk-averse decision maker will adopt the earlier time, i.e. 83h, or, in the initial probability distribution, g^{-1} (83h).

At time 50h, if no failure has taken place yet, the lower bound of the estimated RUL is equal to 37h (and the upper bound 61h), which will lead to an intervention

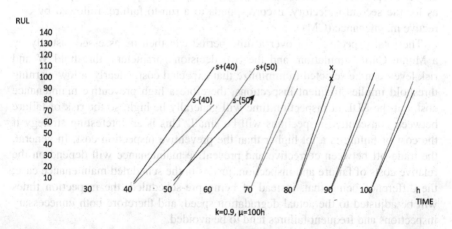

FIGURE 9.4 Predictive maintenance intervals evolving with time. On the y-axis, EOL = end of life= current time+ RUL.

TABLE 9.1
The Successive Values for the Suggested Next Maintenance Intervention Time (under the Assumption That No Failure Occurs until Time *t*)

t (*h*)	0	10	20	30	40	50	60	70	80	90
next PM (h)	68	71	75	79	83	87	91	95	99	102

at 87h. At 60h (if no failure has taken place yet), the suggested maintenance intervention will be at 60h+ RUL− = 60h +31h = 91h and so on.

Table 9.1 shows the successive values for the suggested next maintenance intervention time (under the assumption that no failure occurs until time t).

On that basis, it is possible to build a cost optimization model, similar to the one introduced in Section 9.2 for scheduled maintenance.

One way of doing so (see, e.g., Huynh et al. 2019) is to build a health indicator (a measure of the severity of the degradation) and to define a warning threshold: when the health indicator crosses the threshold, the next preventive inspection is performed after a certain time span, which is defined in such a way that the RUL at the date of threshold crossing has a "small" probability of being less than that time span. "Small" is defined by the degree of risk that the decision maker is willing to take. Therefore the decision when to maintain preventively depends on two parameters: the threshold, and the risk.

This policy is illustrated in Figure 9.5, where two trajectories are shown; when the first trajectory crosses the warning threshold (W), the time to the next preventive maintenance (PM) is calculated (denoted $s*$). In the example in the figure, $s*<RUL$ so indeed the failure will be prevented and the cost incurred will be the cost of the preventive maintenance operation (typically, a preventive replacement); as for the second trajectory, it corresponds to a run-to-failure, followed by corrective maintenance (CM).

The total expected cost over a time period can then be assessed, usually by a Monte Carlo simulation, and the two decision parameters, threshold W and risk level, can be selected to minimize that expected cost. Clearly, a low warning threshold implies frequent inspections, therefore a high preventive maintenance cost, but the RUL at inspection time will generally be high, so the risk of failure between consecutive inspections will be small. This is an interesting strategy if the cost of failure is a lot higher than the preventive inspection cost. In general, the trade-off between corrective and preventive maintenance will depend on the relative costs of failure and inspection, just as in the scheduled maintenance case; the difference being that, instead of being pre-scheduled, the inspection times will be adjusted to the actual degradation speed, and therefore both unnecessary inspections and frequent failures tend to be avoided.

In the above example (Table 9.1), the next inspection time was defined by a 20% risk.

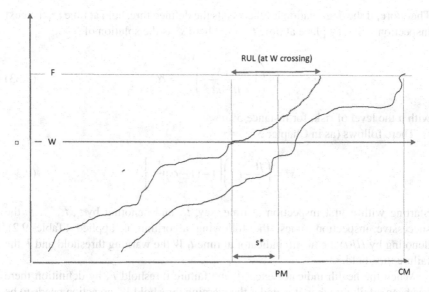

FIGURE 9.5 Two degradation trajectories.

The cost is usually evaluated as a cost per unit of time.
Over a period T:

$$E\left(\frac{\text{Cosr}}{T}\right) = \frac{C(T)}{T},\qquad (9.32)$$

where $C(T)$ is the sum of preventive inspection cost and cost of failures over a period of length T. From now on, a *lower case letter c* will denote the *cost per unit of time*, i.e.

$$c(T) \equiv \frac{C(T)}{T}.$$

An asymptotic unit cost is obtained by letting the duration T of the cycle go to infinity.

If we consider the special family of time-to-failure distributions with MRL linear in time (Chapter 3), into which a time transformation $g(.)$ can convert most distributions (Chapter 5), the RUL distribution at time t (Chapter 3) is given as follows:

$$P\left[\text{RUL}(t) < s\right] = 1 - P\left[\text{RUL}(t) > s\right] = 1 - R_t(s) = 1 - \left(1 - \frac{ks}{\mu - kt}\right)^{\frac{1-k}{k}}.$$

Therefore, if the degradation level exceeds the defined threshold at time t^*, the next inspection will take place at time t^*+s^*, where s^* is the solution of

$$1-\left(1-\frac{ks}{\mu-kt^*}\right)^{\frac{1-k}{k}} = \alpha. \tag{9.33}$$

with α the level of risk, for instance 5%.

There follows (as in Chapter 3):

$$s^* = \left(\frac{\mu}{k}-t^*\right)\left[1-(1-\alpha)^{\frac{k}{1-k}}\right]. \tag{9.34}$$

Starting with a first inspection at time, say, t_0, and denoting by t_1, t_2, ... t_n the successive inspection times, the following algorithm is applied (Table 9.2), denoting by $HI(t)$ the health indicator at time t, W the warning threshold and F the failure threshold.

When the health indicator exceeds the failure threshold F, by definition there has been a failure; when it is under the warning threshold W, no action needs to be taken. In both cases, the next inspection date must be set after some default time τ. When the value of the health indicator is between the warning threshold and the failure threshold, the next inspection date is set according to (9.34).

Once t_{n+1}^* has been found, the corresponding time in the initial distribution is $g^{-1}(t_{n+1}^*)$.

Let us see what cost optimization looks like with this model.

Let N denote the index of the first inspection that finds the asset failed ($HI > F$ at time T_N), and T the time of the first failure. Thus, the asset is assumed to be

---

TABLE 9.2
Pseudo-Code for Predictive Maintenance Algorithm

Pseudo-Code:

$n = 0$
Initialize $t_n^* = t_0$

    IF $HI(t_n^*) > F$

        Corrective maintenance
        Set next inspection date $t_{n+1}^* = t_n + \tau$

    IF: $W < HI(t_n^*) < F$

        Set next inspection date $t_{n+1}^* = s^*$
        from (9.34) with $t^* = t_n^*$

    IF: $HI(t^*) < W$

        Set next inspection date $t_{n+1}^* = t_n + \tau$

$n = n+1$
Continue

renewed preventively (AGAN), for instance replaced, at times $T_1, T_2, \ldots T_{N-1}$ and, at the next inspection, which takes place at time T_N, it is found to have failed.

This model presents a simplification with respect to the algorithm of Table 9.2: the case $HI < W$ (health indicator under the warning threshold) is not considered.

Thus

$$T_{N-1} < T \leq T_N.$$

The cost per unit of time, over the period from the beginning of observations, t_0, to the time T_N, can be expressed as:

$$c(T) = \frac{(N-1)c_1 + c_2}{T_{N-1} + t_f}, \tag{9.35}$$

where t_f denotes the time elapsed between the $(N-1)$st inspection and the failure instant.

$$\left(T_{N-1} \leq T_{N-1} + t_f \leq T_N\right).$$

Indeed, by time T_N, there have been $(N-1)$ inspections that each led to a preventive maintenance operation, with cost c_1, and the Nth inspection revealed a failure (occurred at time $T_{N-1} + t_f$), which induced a corrective maintenance cost, and more generally a failure cost, c_2.

As previously, c_1 denotes the cost of a preventive maintenance operation and c_2, the cost of a failure.

We need the expectation of c(T).

The random variable N has the following probability distribution:

$$P[N = n] = \prod_{j=0}^{j=n-2} R_{t_j}(s_j)\left[1 - R_{t_{n-1}}(s_{n-1})\right] \tag{9.36}$$

which expresses the fact that the $(N-1)$ successive periods: (t_0, T_1), $(T_1, T_2), \ldots \left(T_{N-2}, T_{N-1}\right)$

have been failure-free, and that a failure has occurred during (T_{N-1}, T_N).

But, since the time periods $s_j (j = 0, 2, \ldots n - 1)$ have been defined by (9.34), then, by definition,

$$R_{t_j}(s_j) = 1 - \alpha \quad \text{for all } j = 0, 1, 2, \ldots$$

and therefore N has a geometric probability distribution, i.e. (9.36) is equivalent to:

$$P[N = n] = \alpha(1 - \alpha)^{(n-1)} \quad (n = 1, 2, \ldots) \tag{9.37}$$

and therefore

$$E(N) = \frac{1}{\alpha.}$$ (9.38)

Let us introduce the following notation, for ease of writing:

$$B(k) = 1 - (1-\alpha)^{\frac{k}{1-k}}.$$ (9.39)

Then,

$$t^*_{n+1} = t^*_n + s^*_n = t^*_n + B(k)\left(\frac{\mu}{k} - t^*_n\right) = [1 - B(k)]t^*_n + \frac{\mu}{k}B(k)$$ (9.40)

By applying successively this recursive relation, there follows (for $k \neq 0$)

$$t^*_n = \frac{\mu}{k} - [1 - B(k)]^n \left(\frac{\mu}{k} - t_0\right)$$ (9.41)

(By using the fact that iterative application of the relation

$$x_{n+1} = a\,x_n + b$$

leads to

$x_{n+1} = a^{n+1}x_0 + b\dfrac{a^{n+1} - 1}{a-1}$, and applying it with

$a = 1 - B(k)$

$b = \dfrac{\mu}{k}B(k).)$

Therefore, assuming $t_0 = 0$ without loss of generality,

$$c(T) = \frac{(N-1)c_1 + c_2}{\frac{\mu}{k} - [1 - B(k)]^{N-1}\left(\frac{\mu}{k}\right) + t_f}$$ (9.42)

(for $k \neq 0$).

As $c(T)$, expressed by (9.42), is a function of the risk level α, the value of α can in principle be selected to minimize the expectation of $c(T)$, taking into account the probability distribution of N, given by (9.37). A numerical resolution is needed in general.

As usual, let us start by taking a look at very special limiting cases.

1. For $k \to 0$ (exponential distribution), note, from (9.34)—cf. Chapter 3, that

$$\lim_{k \to 0} s^* = -\mu \log(1-\alpha).$$ (9.43)

Therefore, from (9.40)

$$t^*_{n+1} = t^*_n - \mu \log (1 - \alpha) \quad \text{for all } n = 1,2,\ldots$$

Hence

$$t^*_n = -n\, \mu \log (1 - \alpha)\ n = 1,2,\ldots$$

which actually amounts to scheduled inspections of periodicity $\tau = -\mu \log (1-\alpha)$. For instance for a risk $\alpha = 5\%$, that would be approximately every $\tau = \mu/20$.

$$E\left(t_N{}^*\right) = -E(N)\mu \log(1 - \alpha) = -\frac{\mu \log(1 - \alpha)}{\alpha}$$

For small α (for instance $\alpha \approx 5\%$),

$$E\left(t_N{}^*\right) \approx \mu = \text{MTTF}$$

In fact,

$$\lim_{\alpha \to 0} \frac{\log(1 - \alpha)}{\alpha} = -1. \tag{9.44}$$

Thus on average, the first inspection that reveals a failure takes place after a time MTTF. As the exponential distribution describes a situation of no degradation, i.e. constant MRL, the inspections can only aim at detecting whether the asset is failed (case $HI > F$).

The average cost per unit of time (9.35) is then

$$c(T) = E\left[\frac{(N-1)c_1 + c_2}{t_{N-1} + t_f}\right]. \tag{9.45}$$

As a first approximation, let us replace the random variables by their expectations in (9.45).

The expectation of $(s - t_f)$ is the so-called mean fault detection time or mean undetected fault time, MUFT. It can be shown, from the definition (see e.g. Dersin 2020, Dersin 2012). that:

$$E\left(s - t_f\right) = \frac{s}{1 - e^{-\frac{s}{\mu}}} - \mu. \tag{9.46}$$

Therefore

$$E\left(t_f\right) = s - E\left(s - t_f\right) = s - \left(\frac{s}{1 - e^{-\frac{s}{\mu}}} - \mu\right) = s\left(1 - \frac{1}{1 - e^{-\frac{s}{\mu}}}\right) + \mu$$

where s can be expressed in terms of α through (9.34). There follows:

$$E\left(t_f\right) = \mu\left[1+(1-\alpha)\frac{\log(1-\alpha)}{\alpha}\right]. \tag{9.47}$$

Then, replacing the random variables by their expectations in (9.45) to obtain an approximation of the expected cost per unit of time:

$$E\left[c(T)\right] \approx -\frac{\left(\frac{1}{\alpha}-1\right)c_1+c_2}{\mu\left(\frac{1}{\alpha}-1\right)\log(1-\alpha)+\mu\left[1+(1-\alpha)\frac{\log(1-\alpha)}{\alpha}\right]} \tag{9.48}$$

Taking the limit of $\alpha{\rightarrow}0$, one obtains: $\lim\limits_{t\rightarrow\infty} E[c(T)] = \infty$.

Indeed, permanent inspection leads to an infinite inspection cost per unit of time. On the other hand, for $\alpha{\rightarrow}1$,

$$\mathrm{Lim}\, E\left[c(T)\right] = \frac{c_2}{\mu}$$

which is the average cost of failure per unit of time.

This is of course the preferred policy for an exponential time-to-failure distribution: waiting for the failure, since preventive maintenance has no effect.

2. Let us now consider the other extreme, $k=1$ (Dirac distribution). Then there follows from (9.39) that, if $\alpha{>}0$,

$$\lim\limits_{k\rightarrow1} B(k) = 1$$

And (from 9.40),

$$\lim\limits_{k{\longrightarrow}1} s_1 = \mu$$

and
$s_n = 0$ for $n{>}1$.

Indeed, in that very special limiting case, it is known with certainty (i.e. zero variance) that the next failure will take place at a time μ after the preventive maintenance restoration date.

A maintenance policy could then consist of performing a preventive maintenance operation just before $t=\mu$ so as to avoid the failure.

The average cost per unit of time would then be:

$$c(T) = \frac{c_1}{\mu}.$$

Another policy would be to wait for the failure, in which case the average cost would be:

$$c(T) = \frac{c_2}{\mu}$$

which is what (9.42) gives (with $N=1$), as $t_f = 0$.

Usually $c_2 \gg c_1$ so that the preventive maintenance option is much preferable.

These extreme cases are a way to check model consistency. Let us now look at some more-realistic cases.

3. Uniform distribution: The uniform distribution corresponds to $k = \frac{1}{2}$ (Chapter 3). Then

$$B(\frac{1}{2}) = \alpha.$$

Therefore

$$s_n = \alpha\,(2\mu - t_n).$$

And

$$t_{n+1} = t_n + \alpha(2\mu - t_n) = (1-\alpha)t_n + 2\mu\alpha$$

$$t_{N-1} = 2\mu\left[1 - (1-\alpha)^{N-1}\right]. \tag{9.49}$$

For instance, a 100% risk-prone decision maker ($\alpha=1$) would take $t_n = 2\mu$, i.e. he would inspect at the end of the theoretical asset life time (ignoring the possibility of an earlier failure; that is, in effect, not taking the variance into account). Whereas a totally risk-averse decision maker ($\alpha=0$) would take $t_{n+1}=t_n$, or $s_n = 0$, i.e. he would constantly be inspecting.

In this case of a uniform distribution, one shows easily, denoting t_f the time elapsed since the last inspection t_{N-1}—to be completely rigorous, it ought to be denoted $t_f(N-1)$, that

$$E(t_f) = \mu\alpha(1-\alpha)^{N-1}.$$

Indeed, from (9.49), the time elapsed between the last two inspections is

$$t_N - t_{N-1} = 2\mu\left[1 - (1-\alpha)^{N}\right] - 2\mu\left[1 - (1-\alpha)^{N-1}\right] = 2\mu\left[(1-\alpha)^{N-1} - (1-\alpha)^{N}\right]$$

$$= 2\mu(1-\alpha)^{N-1}\left[1 - (1-\alpha)\right] = 2\mu\alpha(1-\alpha)^{N-1}.$$

And, with a uniform distribution for t_f, $E(t_f) = \dfrac{t_N - t_{N-1}}{2} = \mu\alpha(1-\alpha)^{N-1}$.
The cost can then be expressed as

$$c(T) = \frac{(N-1)c_1 + c_2}{2\mu\left[1-(1-\alpha)^{N-1}\right]+t_f}. \qquad (9.50)$$

It is seen from (9.50) that, if α increases, then, as the distribution of N is shifted toward lower values (the failure occurs earlier), the numerator decreases; but the trend of the denominator is not clear since $(1-\alpha)$ decreases but so does N.

If the random variables N and t_f in (9.50) are replaced with their expectations, the following estimate $\hat{c}(T)$ is derived as a heuristic approximation for the expectation of $c(T)$ obtained by replacing the random variables with their respective expectations in (9.50):

$$\hat{c}(T) = \frac{\left(\dfrac{1}{\alpha}-1\right)c_1 + c_2}{2\mu\left[1-(1-\alpha)^{\frac{1}{\alpha}-1}\right]+\mu\alpha(1-\alpha)^{\frac{1}{\alpha}-1}}. \qquad (9.51)$$

When α increases, the ratio $\hat{c}(T)$ first decreases, then increases, and therefore reaches a minimum for a certain value α^* of the risk α.

From (9.50) or (9.51), it is seen that the optimum risk level is a function of the cost ratio c_2 / c_1 only, not the absolute values of the costs.

For instance, assume $\mu = 100$h. For $\dfrac{c_2}{c_1} = 10$, the minimum is found for $\alpha^* = 0.70$,

as illustrated in Figure 9.6(a). If $\dfrac{c_2}{c_1} = 100$, then the value α which minimizes the expected cost per unit of time becomes $\alpha^* = 0.45$ (Figure 9.6(b)). Indeed, with a much higher failure cost relatively to the preventive replacement cost, there is more incentive to avoid failures through more predictive maintenance: a failure risk of only 45% is tolerated instead of 70% in the previous example.

For a cost ratio $\dfrac{c_2}{c_1}$ of 1000, the optimal risk level would fall to 25%.

It can be verified that calculating the expectation of $c(T)$ from (9.50) with the distributions of N and t_f leads to results that are very close to $\hat{c}(T)$ computed from (9.51).

The inspection maintenance times corresponding to the optimal risk level in the two cases are shown in Table 9.3. It is seen that, for the lower risk level (corresponding to a relatively higher cost of failure), the inspection times occur earlier. (In both cases, they converge to 200h= 2μ, which is the theoretical maximum life time in this example.)

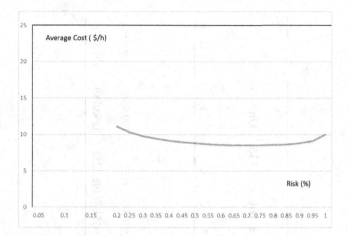

FIGURE 9.6(a) Approximate average maintenance cost per unit of time as a function of risk (uniform distribution example). $(c_2 = 10c_1)$.

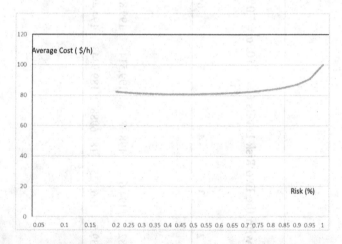

FIGURE 9.6(b) Approximate average maintenance cost per unit of time as a function of risk (uniform distribution example). $(c_2 = 100c_1)$.

These results would therefore also apply to any distribution for which $k = 1/2$, for instance the Gamma distribution with shape parameter $\beta=3$ (see Chapter 5). To convert the times of Table 9.3 into the initial referential, i.e. into the "real world", one has to apply the inverse transformation g^{-1}: if t'_n denotes the nth inspection time in Table 9.3, the real inspection times are given by

$$t_n = g^{-1}\left(t'_n\right), \quad n = 1, 2, 3\ldots \tag{9.52}$$

TABLE 9.3
Predictive Maintenance Times with the Two Risk Levels ($\frac{C_2}{C_1} = 10$, $\alpha = 70\%$ and $\frac{C_2}{C_1} = 100$, $\alpha = 45\%$). Uniform Distribution

n			0	1	2	3	4	5	6	7	8	9	10
$\mu(h)$		100											
$\alpha=07$	0.7	$tn\,(h)$	0	140	182	194.6	198.38	199.514	199.8542	199.9563	199.9869	199.9961	199.9988
$\alpha=0.45$	0.45	$tn\,(h)$	0	90	139.5	166.725	181.6988	189.9343	194.4639	196.9551	198.3253	199.0789	199.4934

The cost for the totally risk-prone policy ($\alpha = 1$) is $\frac{c_2}{\mu}$, entirely dependent on the cost of failures and their frequency, and equal to the ratio of corrective maintenance cost to MTTF (This can be seen already from (9.50) since, in that policy, $N = 1$, there is no preventive inspection).

At the other extreme, in the totally risk-averse policy ($\alpha=0$), the cost is infinite (constant inspection is necessary).

4. Rayleigh distribution

It was shown in Chapter 5 that for this distribution, which is the Weibull distribution with a shape factor β equal to 2, the k parameter is equal to

$$k = \frac{\pi}{2} - 1 \approx 0.57$$

which is close to the value of the k parameter for the Gamma distribution of shape factor equal to 4 ($k = 0.6$ in that case).

The inverse of the time transformation, $g^{-1}(.)$ for the Rayleigh distribution is given (using 5.31 and 5.41) by the following expression:

$$g^{-1}(t) = \eta \frac{4-\pi}{\pi-2} \left[-log \left(1 - \frac{\frac{\pi-2}{2}}{\eta \frac{\sqrt{\pi}}{2}} t \right) \right]^{\frac{1}{2}}, \qquad (9.53)$$

where η denotes the scale parameter of the distribution, so that the expectation μ is given by

$$\mu = \eta \Gamma \left(1 + \frac{1}{2} \right) = \frac{\eta}{2} \Gamma \left(\frac{1}{2} \right) = \frac{\eta}{2} \sqrt{\pi} \approx 0.885 \eta.$$

This will be used to convert the transformed times back to the original referential.

(In the case of the uniform distribution, g(.) is the identity function since the uniform distribution has an MRL linear in time—see Chapter 5).

In Table 9.4, the successive inspection times have been determined, for the Rayleigh distribution, which correspond to the same two risk levels α as considered previously with the uniform distribution (since 0.57 is not too far from 0.50, those values of α should not be far from the cost-optimal one, by continuity).

To that end, (9.34), (9.40), and (9.53) have been used. In order to have the same support $(0, \frac{\mu}{k})$ as previously, the scale parameter has been adjusted, which leads to a value of 128.81h for η, and 114h for the mean μ, instead of 100h in the uniform distribution example.

TABLE 9.4

Predictive Maintenance Times with the Two Risk Levels ($\frac{c_2}{c_1}$ = 10, α = 70%;

and $\frac{c_2}{c_1}$ = 100, α = 45%). Rayleigh Distribution

	Step		0	1	2	3	4	5	6	7	8	9	10
	η (h)	128.81493682											
	β	2											
	mu (h)	114.1592654											
	mu/k(h)	200											
α	70%	$g^{-1}(t)$ (h)	0	106	150	184	213	237	260	281	265	319	336
α	45%	$g^{-1}(t)$ (h)	0	75	106	130	150	168	183	198	231	225	237

Observe that the interval between two successive maintenance inspections diminishes with time: for instance, for a 70% risk level, from 44h for the time between the first two maintenance steps, down to only 17h for the time between steps 9 and step 10.

(Note that, for some examples, μ=100 days is probably more realistic than 100h, but the same observation can be made, it is just a question of scale).

Dynamic Maintenance Policy

Clearly, $s*$, the time to the next preventive inspection, is a function of the "degradation rate" k.

From (9.34), three interesting observations can be made:

1. the closer the last inspection time $t*$ is to the end of life $\frac{\mu}{k}$, the closer the next inspection time will be;
2. the smaller the risk α which the decision-maker is willing to take, the earlier the next inspection time will be (since $s*$ is a decreasing function of α);
3. if k increases, the next inspection will take place sooner. This is to be expected since k measures the rate at which the MRL decreases with time, closely related to the speed of degradation.

Finally, as the actual inspection time is $g^{-1}(t*)$ in the initial referential, the particular shape of the TTF distribution, characterized by the metric dg, will play a role as well.

In reality, due to evolution in context and the onset of new degradation modes, both k and the metric dg may change with time. Statistical estimations, as described

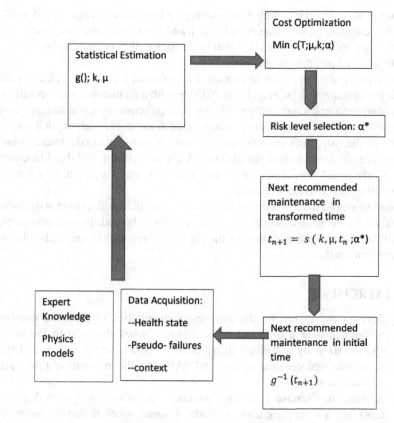

FIGURE 9.7 Conceptual dynamic maintenance decision support flowchart.

in Chapter 8, will be necessary in order to update those parameters dynamically, and thus adjust the predictive maintenance times. This leads to a dynamic updating process which is described in Figure 9.7.

From current assumptions on the g (.) time transformation and the parameters μ and k, an expected maintenance cost minimization is performed, which leads to the selection of an optimal risk level α^*. In general, this step requires Monte Carlo simulation of the cost $c(T)$.

The output of this phase is the recommendation of the next predictive maintenance time, first expressed in the transformed time referential and then, using the inverse of the g transformation, in the initial time referential.

Then come the data acquisition and statistical estimation phases. Data to be acquired—over an updating period to be selected, include failures and pseudo-failures, but also asset health indicators estimated from sensor readings; and external parameters such as environmental conditions or mission profile conditions, in general "context conditions", which may vary with time.

From those elements, and also by drawing on expert knowledge and, if available, physical models, the parameters k and μ, and if necessary the time transformation $g(.)$, are updated, and the algorithm loops back to the cost minimization step, with the updated parameters and assumptions.

Cost minimization need not be explicit. Furthermore, it could be replaced with availability optimization (see e.g. Letot 2013) or with a multi-criteria optimization.

The above examples serve only to illustrate the principles of a maintenance cost optimization model, in a predictive maintenance framework, with explicit consideration of the risks. Any realistic implementation should also take into account maintenance efficiency (which can also be a decision variable), and should include Monte Carlo simulations since analytical treatments quickly reach their limits. This is still a very active research area.

It must be emphasized that the general outcome of such an algorithm is decision support, i.e. the decision of when to maintain is ultimately made by human experts, who rely on the recommendations of the algorithm, rather than automatically by the algorithm itself.

9.5   EXERCISES

1. Give an expression for the probability distribution of $N(t)$, the number of failures over a time span of duration t, when the time to failure is characterized by a Weibull distribution of parameters β and η, and the corrective maintenance is minimal (ABAO); the restoration times are considered negligible.

2. Assume, in Exercise 1, a shape parameter equal to 2 (Rayleigh distribution), and a scale parameter of 1000h. Assume as before that the maintenance is minimal and the restoration times negligible.
   a. What is the expected number of failures over an operating time of 2000h?
   b. What is the probability of fewer than two failures occurring over that period?
   c. What would those numbers be (under a/ and b/ above) if instead, the time to failure were exponentially distributed with an MTTF of 1000h?

3. From the Laplace transform of the uniform distribution over an interval (0, 2μ), Section 9.3, derive the first- and second-order moments, and verify you find the known results: mean μ and variance $\dfrac{\mu^2}{3}$. (Hint: See Chapter 3, moment generating function).

4. Same exercise for the exponential distribution.

5. Same exercise for the Dirac distribution (check that you find $\sigma^2=0$).

6. Prove (9.46).

7. Calculate the expectation of the cost per unit of time from (9.50), using the fact that N has a geometric distribution, and $t_f(N)$ a uniform one, given N; and compare with the approximate expression given in (9.51).

8. In the cost model presented in Section 9.4, study the cost-per-unit-of-time function for $k > \frac{1}{2}$. Compare with the uniform distribution case ($k = \frac{1}{2}$). Show that the inspections should occur sooner to keep the same risk.

BIBLIOGRAPHY

Alessi, A., La Cascia, P., Lamoureux, B., Pugnaloni, M., & Dersin, P., " Health Assessment of Railway Point Systems: A Case Study", Proceedings of the Prognostics & Health Management Society Conference, Bilbao, Vol. 3, 2016.

Asset Management Excellence: Optimizing Equipment Life-Cycle Decisions, 2d Edition, Edited by John D. Campbell, Andrew K.S. Jardine, Joel McGlynn, Boca Raton, FL, Taylor & Francis, CRC Press, 2019.

Atamuradov, V., Medjaher, K., Camci, F., Zerhouni, N., Dersin, P., & Lamoureux, B., "Machine Health Indicator Construction Framework for Failure Diagnostics and Prognostics", *Journal of Signal Processing Systems*, 92, pp. 591–609, 2020.

Birolini, A., *Reliability Engineering: Theory and Practice*, 8th Ed., Springer, New York, 2017.

Dersin, P., & Péronne, A., "Probabilistic Characterization of Undetected Fault Time in a Redundant System with Imperfect Detection", Proc. Lambda-Mu Symposium, Lambda-Mu 18, 2012.

Dersin, P., & Thévenet, Q., "The Internet of Rail, Smart Maintenance of Passenger Access Doors for Tomorrow's Rail Mass Transit", UITP Congress, Stockholm, 2019.

Dersin, P., & Valenzuela, R., "Designing for Availability in Systems, and Systems of Systems", Tutorial, "67th Reliability & Maintainability Symposium" (RAMS 2020), Tucson, AZ, 2020.

Doyen, L., & Gaudoin, O., "Classes of Imperfect Repair Models Based on Reduction of Failure Intensity or Virtual Age", *Reliability Engineering and System Safety*, 84, pp. 45–46, April 2002.

Finkelstein, M., *Failure Rate Modelling for Reliability and Risk*. London, Springer, 2008.

Finkelstein, M., "On the Optimal Degree of Imperfect Repair", *Reliability Engineering and System Safety*, 138, pp. 54–58, 2015.

Huynh, K., Grall, A., & Bérenguer,C. "A Parametric Predictive Maintenance Decision-Making Framework Considering Improved System Health Prognosis Precision", *IEEE Transactions on Reliability*, 2019, 68(1), pp. 375–396. 10.1109/TR.2018.2829771. hal-01887

IEC 60030-6-14, Dependability Management. Part 3–14: Application Guide – Maintenance and Maintenance Support, 2004.

Kijima, M., "Some Results for Repairable Systems with General Repair", *Journal of Applied Probability*, 26, pp. 89–102, 1989.

Letot, C., "Predictive Maintenance of Industrial Assets Based on Modelling, Estimation and Simulation of Degradation Laws" Ph.D. Thesis, University of Mons, Belgium, 2013 (in French).

Nachlas, J., *Reliability Engineering Probabilistic Models and Maintenance Methods*. Boca Raton, FL, Taylor & Francis, CRC, 2005.

Ross, S.D., *Introduction to Probability Models*, 10th Ed., Burlington, MA, Academic Press(Elsevier), 2010.

Selby, S.M, "Standard Mathematical Tables", 12th Edition, Cleveland, OH, The Chemical Rubber Co., 1972.

Wang, H., & Pham, H., "A Quasi-Renewal Process and Its Applications in Imperfect Maintenance", International Journal of Systems Science, 27, pp. 1055–1062, 1996.

10 Advanced Topics and Further Research

10.1 THE GINI INDEX

The Gini coefficient (Sen 1997), which is used by economists in order to measure income inequality in a population, has been introduced recently into reliability theory (Kaminsky 2008, Krivtsov 2019.).

This index is comprised between −1 and +1.

For non-repairable systems, its value is nonnegative: strictly positive for increasing failure rate distributions, and equal to 0 for the exponential distribution.

The Gini index $G(T)$, generally a function of time, is defined by reference to the exponential function.

It is built from the cumulative hazard function $\Lambda(T)$ (see Chapter 2).

Considering the time interval $(0,T)$; if the distribution was exponential, with rate λ, its cumulative hazard function would be equal to λT.

We consider the particular exponential distribution which is such that its cumulative hazard rate at time T coincides with that of the considered function. Therefore

$$\lambda(T) = \frac{\Lambda(T)}{T}. \tag{10.1}$$

Then, to define the Gini coefficient at time T, the integral of the cumulative hazard function over $(0,T)$ is compared to that of that particular exponential distribution, which is $\dfrac{T}{2}\Lambda(T)$.

Formally the Gini index at time T is defined by

$$G(T) = 1 - \frac{2\int_0^T \Lambda(t)\,dt}{T\Lambda(T)}. \tag{10.2}$$

By construction, $G(T)$ is identically 0 for an exponential function.

Since the coefficient k, the slope of the mean residual life (MRL) in the family of distributions studied earlier (introduced in Chapter 3) also measures the

DOI: 10.1201/9781003250685-10

"inequality" of the distribution, i.e. its departure from the exponential distribution, it is interesting to examine its relation to the Gini index.

Therefore we derive here an expression for the Gini index, as a function of k, for the special family of distributions with MRL linear in time.

For that family, recall (Chapter 3) that the failure rate is expressed as follows:

$$\lambda(t) = \begin{cases} \dfrac{1-k}{\mu - kt} & 0 \leq t \leq \dfrac{\mu}{k} \\[4mm] 0 & t > \dfrac{\mu}{k} \end{cases}. \tag{10.3}$$

And accordingly the cumulative hazard function is given by

$$\Lambda(t) = -\frac{1-k}{k} \log\left(1 - \frac{kt}{\mu}\right). \tag{10.4}$$

Therefore, application of the definition (10.2) of the Gini index yields

$$G(T;k) = -1 + 2\frac{\mu}{kT} + \frac{2}{\log\left(1 - \dfrac{kT}{\mu}\right)}. \tag{10.5}$$

From (10.5), various properties can be verified.

i.   Exponential distribution: $k= 0$. $\lim\limits_{k \to 0} G(T;k) = 0$ as expected

To see this, let $x = \dfrac{kT}{\mu}$.

Then $0 < x < 1$ and

$$G(T,k) = G(x) = \frac{2}{x} + \frac{2}{\log(1-x)} - 1. \tag{10.6}$$

Using the Taylor–McLaurin series expansion of $\log(1-x)$, the following property can also be proved.

ii. $\lim\limits_{T \to \frac{\mu}{k}} G(T;k) = 1$ whatever the value of k.

For the Dirac distribution ($k = 1$) or the uniform distribution ($k = 1/2$), the Gini index does not have any remarkable value.

The dependence of the Gini index on k is illustrated in Figure 10.1. It is an increasing, convex function of k.

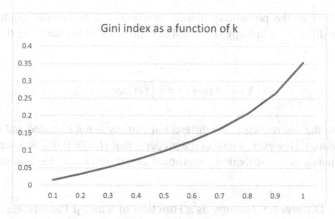

FIGURE 10.1 Gini index as a function of k parameter.

10.2 ENTROPY AND THE K PARAMETER

10.2.1 ENTROPY

As is well known, the entropy $H(p)$ of a discrete random variable X with probability distribution $\{p_i\}$, introduced by Shannon in the context of information theory (Shannon 1948), is a number defined as follows:

$$H(p) = -E[\text{Log} X] = -\sum_{i=1}^{i=n} p_i \log p_i. \tag{10.7}$$

It is a number between 0 and 1 that measures the expectation of the information acquired when learning the outcome of a random experiment. For instance, if the probability distribution is deterministic, i.e. $p_j = 1$ and $p_i = 0$ for $i \neq j$, then $H(p) = 0$, which is consistent with the fact that the event (the value taken by X) was known a priori with certainty, therefore its realization brought no information.

On the other hand, the entropy is maximum if $p_i = \dfrac{1}{n}$ for all i. In that case, $H(p) = \log n$.

Since the family of distributions introduced here (those with an MRL linear with time) is characterized by a parameter k which measures the degree of uncertainty (maximal for $k = 0$ and minimal for $k = 1$), it is natural to relate that parameter to the entropy of the distribution.

But, to do so, we need to use the concept of differential entropy, which applies to continuous distributions.

10.2.2 DIFFERENTIAL ENTROPY

Differential entropy is the generalization of the concept of entropy defined by (10.7), to continuous probability distributions.

If $f(t)$ denotes the probability density function of a continuous random variable X, the differential entropy is defined as follows (with log denoting the natural logarithm):

$$H(X) = -E[\log f] = -\int_{-\infty}^{+\infty} f(t) \log f(t) dt. \tag{10.8}$$

Contrary to the discrete case, the differential entropy is not bounded and can take negative values. However, it retains the property that $H(X) = 0$ if X is deterministic and that, in a class of distributions, maximum uncertainty coincides with maximum entropy.

10.2.2.1 Differential Entropy as a Function of k and μ Parameters

For the family of TTF distributions with an MRL linear with time, application of (10.8) yields the following expression for the differential entropy.

$$H(k,\mu) = -\int_0^{\infty} f(t) \log f(t) dt \tag{10.9}$$

where (see Chapters 3 and 8) the density $f(t)$ is given by

$$f(t;k,\mu) = \begin{cases} \dfrac{1-k}{\mu}\left(1-\dfrac{kt}{\mu}\right)^{\frac{1}{k}-2} & 0 \leq t \leq \dfrac{\mu}{k} \\[2mm] 0 & \dfrac{\mu}{k} < t \end{cases} . \tag{10.10}$$

From (10.9) and (10.10), the entropy can be calculated, which results in the following expression:

$$H(k,\mu) = \log\mu + \frac{1-2k}{1-k} - \log(1-k). \tag{10.11}$$

The differential entropy is plotted as a function of k in Figure 10.2. The mean μ is just a shift parameter. The function remains relatively constant for values of k smaller than 0.9, after which it drops abruptly.

The following special cases can be verified:

i.   For $k = 0$ (exponential distribution),

$$H(0, \mu) = 1 + \log \mu$$

which is the maximum value of the entropy function $H(k,\mu)$

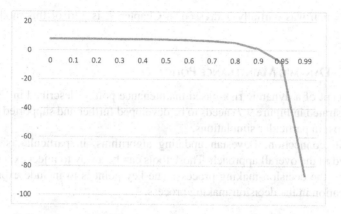

FIGURE 10.2 Differential entropy as a function of the k parameter.

ii. For $k \to 1$ (Dirac distribution),

$$\text{Lim } H(k,\mu) = -\infty$$

which is obviously the minimum value.

iii. For $k = \frac{1}{2}$ (uniform distribution), an intermediate value is obtained:

$$H(\tfrac{1}{2}, \mu) = \log(2\mu) = \log(2) + \log \mu < 1 + \log \mu.$$

Thus the behavior of the entropy function as a function of k is somewhat similar to that of the variance, except that it is not bounded.

10.3 PERSPECTIVES FOR FUTURE RESEARCH

A novel approach has been presented and illustrated, which resides at the crossroads between reliability engineering and PHM, and centers on the notion of RUL dynamics modeling, including uncertainty and "RUL loss rate", and the tool provided by the "time warping".

We hope that useful new perspectives have been offered and that, at the same time, fertile research activities can emerge in order to deepen and broaden the results presented in this book.

Let us outline some perspectives.

10.3.1 COMPLEX SYSTEMS

Complex redundant structures behave differently from simple components; including a description of the redundancy mechanisms in the formalism introduced here is probably a useful area of investigation. The subject of multiple degradation modes, which was partially addressed in Chapter 7, is part of this modeling of complexity.

10.3.2   DYNAMIC MAINTENANCE POLICY

The concept of a dynamic risk-based maintenance policy, described in Chapter 9 (and illustrated in Figure 9.7) needs to be developed further and supported by more algorithms, in particular simulations.

In that connection, Bayesian updating algorithms, in particular, should be integrated in the overall approach. Those tools can be a way to inject expert knowledge into the decision-making process. The key point is to include explicit risk quantification in the decision-making process.

10.3.3   SIGNAL PROCESSING

The relation between the "average RUL loss" and the coefficient of variation seems to us to be a key result. In signal processing, one of the definitions of the signal-to-noise ratio is the reciprocal of the coefficient of variation. At one point (Chapter 7), it was shown that introducing even a minute amount of randomness (i.e. increasing noise) in the degradation processes could have a substantial impact on the RUL dynamics qualitative behavior. Further investigation of those considerations might contribute to enhancing synergies between signal processing and PHM.

10.3.4   PHYSICS AND MACHINE LEARNING

This book has purposely not addressed machine learning. It is realized that, as stated in the introduction, that field—and in particular neural network-based "deep learning", holds tremendous promises (Fink 2020), and could probably be combined with the approach described here. For instance, direct estimation of the g warping transformation from field data might perhaps be achieved with deep-learning methods. The recent successful applications of variational auto-encoders and physics-informed neural networks (Arias 2021) make one hopeful about that possibility.

Recent applications (Li 2022) have shown also that deep survival models (i.e. deep learning combined with survival functions, aka reliability functions) could be a powerful tool.

The above list is of course not exhaustive. It is expected that, as more and more cross-fertilization takes place between theory and applications, it will grow continually.

BIBLIOGRAPHY

Arias-Chao, M., "Combining Deep Learning and Physics-Based Performance Models for Diagnostics and Prognostics", Ph.D. Thesis, ETH-Zürich, 2021.

Fink, O., Wang, Q., Svensén, M., Dersin, P., Lee, W-J., & Ducoffe, M., "Potential, Challenges and Future Directions for Deep Learning in Prognostics and Health Management Applications", *Engineering Applications of Artificial Intelligence*, 92, 2020, 103678, ISSN 0952-1976.

Kaminsky, M., & Krivtsov, V., "An Integral Measure of Aging/Rejuvenation for Repairable and Non–repairable Systems", *Reliability and Risk Analysis: Theory & Applications*, 1, pp. 69–76, 2008.

Krivtsov, V., & Kaminsky, M. "Recent Advances in Reliability Applications of Gini Type Index," *2021 Annual Reliability and Maintainability Symposium (RAMS)*, 2021, pp. 1–4, doi:10.1109/RAMS48097.2021.9605771

Li, X., Krivtsov, V., & Arora, K., "Attention-Based Deep Survival Model for Time-Series Data", *Reliability & System Safety*, 217, 2022, 108033, ISSN 0951-8320.

Sen, A., *On Economic Inequality*. Oxford, Clarendon Press, 1997.

Shannon, C.E., "The Mathematical Theory of Communication", *Bell System Technical Journal*, 27, pp. 370–423; pp. 623–756, 1948.

Index

Note: Page numbers in **bold** refer to tables and those in *italic* refer to figures.

Printed in the United States
by Baker & Taylor Publisher Services

Printed in the United States
by Baker & Taylor Publisher Services